DETOX

Also by Merla Zellerbach:
THE TYPE 1/TYPE 2 ALLERGY RELIEF PROGRAM
(with Alan Scott Levin, M.D.)
LOVE IN A DARK HOUSE

DETOX

*A Successful & Supportive Program for
Freeing Your Body from the Physical
and Psychological Effects of*

• CHEMICAL POLLUTANTS (At Home & At Work) •

• JUNK FOOD ADDITIVES • SUGAR •

• NICOTINE • DRUGS • ALCOHOL • CAFFEINE •

• PRESCRIPTION AND

NONPRESCRIPTION MEDICATIONS •

• AND OTHER ENVIRONMENTAL TOXINS •

Phyllis Saifer, M.D., M.P.H.
Merla Zellerbach

JEREMY P. TARCHER, INC.
Los Angeles
Distributed by Houghton Mifflin Company
Boston

Library of Congress Cataloging in Publication Data

Saifer, Phyllis.
    Detox: a successful and supportive program for freeing your body from
the physical and psychological effects of chemical pollutants at home and
work, junk food additives, sugar, nicotine, drugs, alcohol, caffeine,
prescription and nonprescription medication, and other environmental
toxins.

    Bibliography: p. 211
    Includes index.
    1. Substance abuse—Treatment. 2. Food, Junk.
3. Food additives—Toxicology. I. Zellerbach, Merla.
II. Title.
RC564.S23   1984   613   84–8491
ISBN 0–87477–332–6

Jeremy P. Tarcher, Inc.
9110 Sunset Blvd.
Los Angeles CA 90069

Design by Tanya Maiboroda

Manufactured in the United States of America
S 10 9 8 7 6 5 4 3 2 1
First Edition

For Fred, Gary, and Linda; Mark, Scott, and Landi; and all the patients who so graciously shared their experiences.

ACKNOWLEDGMENTS

Very special thanks to our editors Janice Gallagher and Robin Raphaelian; to Beth and Dr. Sandor Burstein, Dr. John Epstein, Zelda Foreman, Dr. Diane L. Hambrick, Fred Hill, Laura E. Jones of CareUnit, Dev Kettner, Dr. John Michael, Dr. Kevin Olden, Dr. Doris Rapp, Dr. Jeffrey Rochford, Lottie Sugarman, and R. Michael Wisner of Detox. Our special gratitude to Dr. Theron Randolph, Dr. William Rea, Dr. Larry Dickey, and Dr. John Maclennan, for pointing the way, and to Dr. Alan Levin for energetic input.

Clinical histories are based on actual patients, but names and details have been changed to protect privacy.

Contents

Introduction

It is now a fact that harmful substances are everywhere: in the air we breathe, the water we drink, the fresh vegetables we eat, and the clothes we wear. The environment, once so familiar and trustworthy, is becoming a stranger as toxic chemicals permeate our atmosphere, lakes, oceans, and soil.

In recent years, doctors, scientists, and nutritionists have begun to realize that common physical and mental complaints, ranging in magnitude from chronic headaches and irritability to immune system deficiencies and depression, are often the result of an accumulation of these toxic substances in the body. The enormous strides made in symptom diagnosis have also found that the causes of toxicity are increasing. A toxin may be as common as the nicotine in a cigarette and the caffeine in a cup of coffee or as subtle as the traces of pesticide in tobacco crops and the gas used to roast the coffee beans. Chemical contamination is so widespread that simply cutting down on food additives is no longer enough to maintain good health.

This book presents an original, step-by-step General Detoxification Program that will safely reduce or eliminate these toxins in your body and your environment. Along with this program, we offer a series of medically tested procedures that will help you withdraw, physiologically and emotionally, from habit-forming toxins such as alcohol, recreational drugs, and medication. Both the general program and the individual procedures implement the most current information available from scientific research along with case histories from Dr. Saifer's seventeen years of private practice.

While we are convinced that eliminating toxins from your body is one of the safest and most effective roads to lasting fitness, we must acknowledge that there are limits to any detoxification process. It won't cure cancer and it won't mend a fracture. It can, however, guide you to a life-style that will support and augment medical treatment and possibly make future care unnecessary. Your days of passive acceptance are over; you can immediately start to improve and safeguard your own health, and the health of your loved ones.

1

The Toxic Reaction

Doctors are often amazed at the number of people who tolerate or ignore mildly unpleasant symptoms. These individuals think that everyone is drowsy after meals, wakes up with a slight headache, or feels claustrophobic around smokers. Because their suffering is neither severe nor disabling, they shrug off minor ailments as too insignificant to treat. Afraid of being called complainers or hypochondriacs, they make little or no effort to seek treatment for or to determine the source of their symptoms. The result is a large population of semihealthy citizens suffering a variety of avoidable illnesses from a myriad of avoidable causes.

These causes may be attributed to one or more of the various toxins—substances that chemically react in the body and produce damage—abounding in the environment. Our bodies recognize these enemies and send us distress signals in the form of headaches, rashes, nausea, chronic fatigue, joint aches, and other symptoms. Inner alarms may seem insignificant in our busy lives, but they signify a "breakdown" in the body and should not be ignored—particularly when there is an effective means of alleviating them.

The concept of self-purification was part of religious cult dogma two decades ago; today, it is medically validated and publicly accepted. This acceptance is due in part to media coverage, which regularly features stories of widespread toxic contamination. Barely a day passes without news of another polluted river, chemical spill, or abandoned dump site leaching poisons into our air and water. These incidents are newsworthy, inspiring great waves of emotion and controversy. What is not widely publicized is our steady exposure to tens of thousands of toxic substances we may never have considered as injurious.

Recent scientific research has shown that both natural and synthetic chemicals directly affect our well-being; for instance, many "hopeless" mental disorders such as depression and schizophrenia have been magically "cured" by the simple removal of a food or one of its ingredients. "Untreatable" rashes suddenly disappear when the sufferer stops taking aspirin, moves to a different area, or changes hairdressers and escapes a toxic spray or shampoo.

GOOD NEWS FOR YOUR BODY

Hearing about or experiencing the effects of toxic exposure make us "bodywise"—concerned about what we take into our systems in the form of food, air, and chemicals. Citizens of the 1980s know more about health and environmental toxins than any previous generation. Each individual wants a body that is free of irritants and poisons, and assurance it will stay that way.

This assurance is nearly guaranteed as medical and scientific advances now make it possible to "detox"—that is, reverse many common manifestations of toxicity. The procedure entails reducing or eliminating the toxins; unless the damage is too advanced to be reversed, your health will greatly improve. The picture painted of us as victims of continual toxic assault need not be a true one. We are far from helpless. We have the means to control this aspect of our lives.

ARE YOU TOXIC OR ALLERGIC?

To start the detox program, you first need to determine whether your reactions are due to toxicity, to allergy, or to other possible causes. The check list at the end of this chapter is designed to help you identify symptoms that are likely to be due to toxicity.

An allergic reaction is an abnormal response to a substance that does not cause symptoms in most people. Many allergic responses are inherited. You can be allergic to your dog, the grass in your yard, or even—in rare and unusual cases—your own blood. These substances are not toxins.

A toxic reaction is a universal—not an individually reactive—response to a poisonous substance. The head- and stomachaches an entire neighborhood feels after the terrain has been sprayed with herbicide are normal physiological responses. Amanita mushrooms and rattlesnake bites harm everybody. These are clear and well-defined toxins.

One of the easiest ways to decide whether your symptoms are caused by a toxin or an allergen is to ask yourself, Am I the only one who is ill?

Gene, for example, has allergies to pollens, dust, molds and animal dander. His wife, Sue, is not allergic. One day they drive to the park for a company picnic. Almost immediately Gene's nose starts to run; his individual biochemistry has made him sensitive to the grass. Within twenty minutes, however, Gene, Sue, and all the other people in the park begin to feel malaise and nausea, and so they pack up their baskets and leave. Later, they find out the tree trunks had been painted with a potent rat killer a few hours prior to their coming. The would-be picnickers all had *toxic* reactions to a pesticide; Gene alone had an *allergic* response to the pollens.

Laura, on the other hand, suffers allergies to chemicals in the environment. Pollens and dust don't bother her, but a whiff of exhaust fumes from a car, the scent of perfume— or of almost any synthetic compound—makes her feel confused, giddy, or depressed. Like Gene and Sue, Laura went to the park to relax in the fresh air, but her chemical allergies made her react to the pesticide much sooner than the others and much more severely. The rat killer was toxic to everybody, but it was both a toxin and an allergen (a substance that causes allergies) to Laura, so her symptoms were compounded.

A case from Dr. Saifer's files records the history of two boys who went to the dentist for fillings. The older brother had no reaction, but that evening, the younger child suddenly began throwing tantrums and crying for no apparent reason. His mother, a long-time patient of Dr. Saifer's, suspected allergies and brought him in for tests. After determined sleuthing, the boy's hyperactive behavior was traced to the plastic in his new inlay. When the dentist replaced it with gold, the youngster reverted to normal.

In this instance, the plastic was an allergen, but only for the one brother who had developed a sensitivity. If the same plastic had somehow gotten into a public drinking water system in large enough amounts to make all recipients sick, it would then be considered a toxin.

The line of demarcation between allergy and toxicity is not always 100 percent clear. Doctors who see children with mysterious skin rashes or headaches often mistakenly diagnose them as allergic. They prescribe antihistamines or steroids but the youngsters do not improve. Sometimes the family of a sick child will move to a new climate to escape the supposed allergens. When the child recovers, it is for a different reason: he is no longer playing in an area built on top of an abandoned factory, or on landfill seeping with chemical residue, and his body has had a chance to detoxify.

A toxin can be as insidious as the invisible asbestos dust seeping from the walls of your old apartment building, as seemingly harmless as a pitcher of cream from cows who nibble on pesticide-contaminated grass, or as obvious as radiation from an X ray. Although bacteria and viruses also fit the definition, they are more difficult to control and do not necessarily respond to detox treatment. Your general health and habits will benefit from the program, but a case of viral herpes, for example, would not be cured.

The main clue to toxicity is that others feel as you do and react at more or less the same time. If you are working in an office and everyone starts to feel nauseated or headachy toward late morning, you are probably all responding to a toxin. It could be fumes in the ventilation system, a potent disinfectant the janitor uses every day at that hour, or the outgassing—outpouring of fumes—from the chemical insulation in the walls.

Symptoms that you alone experience are most likely allergy-related. If you sneeze when you vacuum a dusty basement, break out in hives after eating strawberries, or feel "spacy" while breathing the exhaust fumes of heavy traffic, suspect allergy. If you always get a headache when you use the office copier—and no one else in the office seems to be bothered—again, allergy is the logical diagnosis. (See *The Type 1/Type 2 Allergy Relief Program* by Levin and Zellerbach for more detailed examples.)

Allergic reactions may have a seasonal pattern and be better or worse at certain times of the year; toxic effects, on the other hand, are often steady, gradual, and cumulative. Obvious symptoms such as sneezing fits are quickly diagnosed as allergies, whereas ailments such as swollen glands, hair loss in clumps, changes in skin pigmentation, or teary eyes without accompanying nasal secretions are usually ascribed to toxins.

Many toxic substances are also carcinogenic. They have been proven to cause cancer, often in infinitesimal doses, in laboratory animals. The difference between toxicity and carcinogenicity is mainly one of time. *Toxins produce symptoms fairly promptly, whereas it may take years of repeated exposure before a cancer appears.* For this reason, epidemiologists—medical specialists who study epidemics or frequent occurrences of diseases—require time to analyze their research and make connections; for example, there is usually a twenty- to thirty-year time lag between a person's first pack of cigarettes or first inhalation of asbestos fibers, and the first signs of cancer. It could take an equally long time to link cancer with toxic substances in common use today.

Toxins have also been known to trigger *autoimmune*

reactions in which the body reacts against its own tissue, resulting in diseases such as systemic lupus erythematosus, multiple sclerosis, thyroiditis, and rheumatoid arthritis. Most scientists now believe that repeated, untreated toxic exposures will inevitably lead to some kind of serious illness.

Whether a substance is a toxin or an allergen, if it is making you ill, see your doctor. If you know what the offender is, eliminate it as soon as possible. Toxins are often more obvious and more within our control and awareness than allergens.

WHO IS SUSCEPTIBLE?

While toxins are universally injurious, reactions to them are neither uniform nor predictable. An unrepaired gas leak in the basement could give you dizzy spells, cause your neighbor to have aching joints, and leave your child unable to concentrate on his homework.

Seven factors determine your response to a given toxin and the degree of damage it can produce.

How much of the toxin is ingested, inhaled or absorbed. Getting twenty chest X rays exposes you to twice as much radiation as ten. You can breathe polluted air for thirty minutes with no ill effects, but after a two-hour walk, you may begin to get symptoms.

The potency of the substance. Dioxin, a defoliant and industrial waste product, is the most potent carcinogen known. The state of New York considers fish with more than 10 parts per trillion of dioxin unsafe to eat. By contrast, it takes 300 million parts per trillion of PCBs (polychlorinated biphenyls, a chemical used to retard heat buildups in electric equipment) to contaminate the air to an unacceptable level.

Frequency of contact. The most common exposures to toxins are on the job, especially in offices, factories, laboratories, and farms; accidental poisoning in the home; adverse reactions to medication or to ordinary products such as cleaning fluid; proximity to industrial wastes, pesticides, or bad air; and abusive use of drugs for behavioral or psychic effects.

Inherited susceptibilities. You may be physiologically or psychologically predisposed to react to certain substances. If one or both parents had emphysema, lung cancer, or asbestosis, for instance, your lungs could be more vulnerable to inhalants.

Current state of health. An accident, surgery, or viral illness in the recent past can weaken your defenses (especially your liver and immune system) and make you more prone to react.

Other stresses. Worry, grief, anxiety, and all negative emotions reduce the body's natural defenses and heighten vulnerability.

Tolerance threshold, also called the "full barrel" or "overload phenomenon." Everyone has a specific level, or tolerance threshold, which should not be exceeded. If the amount of toxins coming into the body stays under that level, the system will adapt and metabolize or excrete the poisons harmlessly. Once the toxins reach that level, however, the tiniest insult in the form of food additives, air pollutants, drugs, or other chemicals will overload the system and cause it to overflow in the form of mental or physical symptoms.

In 1953, Ambassador Clare Booth Luce moved into the American Embassy in Rome. She had no health problems for a year, then began to feel vaguely tired and ill. Attacks of tension and nausea followed. Doctors diagnosed anemia and nervous fatigue.

The symptoms grew worse. Her hair came out in handfuls, her fingernails became brittle, her teeth began loosening. Finally, one physician suspected arsenic poisoning; blood tests confirmed it. Within a week the arsenic was traced to the paint on her bedroom ceiling. It had sifted down in tiny amounts of dust, dissolving in her breakfast coffee. The room was redone with arsenic-free paint, and after a period of recovery, Mrs. Luce returned to her duties.

It took twelve months for the ambassador to become ill; it might have taken another person six or eighteen months. If Mrs. Luce had spent more time in the room, doubled her coffee intake, or had recently had surgery or a viral illness, she might have reacted sooner.

Another case concerns a twenty-two-year-old medical student who was exposed to large doses of formaldehyde in his anatomy lab. He began to have headaches and rashes but felt better the moment he left class. After continued exposure he started to get the same reactions at home to small amounts of formaldehyde in his toothpaste and shampoo. By eliminating as much exposure as possible and by following the general detoxification program described in Chapter 4, he was able to lower his chemical load and regain his health.

WHAT YOUR SYMPTOMS MEAN

If you are puzzled about the source of your health problems, take a moment to evaluate your condition and consider some of the possibilities. Certain symptoms are more commonly associated with toxic reactions than others. We are not including the acute physical trauma of swallowing strychnine or inhaling nerve gas, or the sudden fever, vomiting, diarrhea, and sunburn-like rash with peeling of the palms and soles that specifically indicate toxic shock syndrome (TSS). This severe illness almost exclusively affects menstruating women using tampons. Both TSS and acute poisoning require different approaches and treatment.

Our concern is with more subtle symptoms. They may be gradual or recurring, from short-term exposure, or they may indicate *chronic systemic toxicity*—damage to skin or internal organs from repeated or long-term exposure. Because these symptoms defy diagnosis, they are often tolerated or ignored. Many could be remedied easily by removing the causative substance, if it were known. The clues we supply should help you determine whether or not you are a victim of toxicity.

The following head-to-toe checklist, which indicates what may—or may not—be a response to a known toxin, should be used in conjunction with your personal and family medical history.

Symptoms of Toxicity

Likely to be Environmentally Induced	*Unlikely*
Head	
Headache that comes on at the same time every day and is experienced by others in the area	Headache that comes on irregularly and in times of tension, anger, or other emotional trauma
Migraine, usually moderate, generalized, pulsating, and constant and not associated with stress nor previously experienced; often related to alcoholism, morphine in medication, or carbon monoxide from car exhaust	Migraine that is severe, throbbing, abrupt, one-sided, or with visual changes; suspect food allergy, stress, brain or other disease
Muscle tension headache, moderate, feels tight or stiff; often from breathing polluted air	Headache with fever and facial numbness; may be stroke, or infection
Sudden hair loss or hair coming out in clumps; often traced to factory fumes, radiation, cancer chemotherapy, or strong beauty products	Gradual hair thinning; often due to genes or age

Likely to be Environmentally Induced	*Unlikely*
Brain	
General restlessness and irritability in any or several specific locations from no ostensible cause; often traced to chemical fumes, fabrics, etc.	Emotional distress due to specific worry or event
Sudden mood or behavior changes on exposure to synthetic chemicals; often from inhaling manganese dust from batteries	Emotional reaction that may be traced to physical or mental trauma
Untimely sleepiness or insomnia; often associated with drug use	Insomnia due to stress or physical pain, discomfort
Tension and jumpy nerves; often from caffeine or drugs that contain caffeine	
Difficulty thinking or concentrating; often associated with lead intoxication from toys, rubber, newsprint, paint	Inability to concentrate with headache, vomiting, blocked sinuses; may be related to brain disorder
Inexplicable depression and burnout that disappear on weekends and vacations; often related to job toxins, including radiation	

Likely to be Environmentally Induced	*Unlikely*
Dizziness; often from alcohol, antibiotics, opiates	Sudden attacks of vertigo with deafness, nausea; may indicate Meniere's syndrome
Loss of appetite with heightened stimulation, irritability, dry mouth, chest pain; often traced to amphetamines, decongestants	Loss of appetite from depression or other causes
Diminished libido (sex drive); often from exposure to vinyl chloride and other industrial solvents; often associated with drug use	Diminished libido from disease or personal problems

Eyes

Eyelid tremor; often the first sign of mercury vapor poisoning in factories	Loss of peripheral vision and halos around electric lights; associated with glaucoma or soft contact lenses
Stinging; often from ozone in office copiers and smog	Progressive, painless loss of vision; associated with cataracts, age, or improper glasses
Teary eyes; often from fumes in energy-sealed buildings	Pain in one eye only due to infection
Irritation or redness, often from cosmetic ingredients	

Likely to be Environmentally Induced	*Unlikely*
Ears	
Temporary deafness and tinnitus (ringing); often from drug use, mainly aspirin	Tinnitus from prolonged exposure to loud sounds such as rock music or drills
Short-term dizziness or tinnitus; often associated with carbon monoxide, antibiotics, diuretics, alcohol, heavy metals	Itching, pain, and foul-smelling discharge from infection
Nose	
Itchy nose and palate; often from food additives, polluted air	Thick yellowish discharge indicating infection
Inability to smell; often from chemicals in photographic supplies, etc.	Loss of smell with headache; possibly linked to neurologic disease
Frequent nosebleeds; often from industrial fumes and herbicides	Nosebleeds with crusting; often from staphylococcal infection, cold dry air, high altitudes, or hard blowing
Mouth	
Toothache, swollen jaw, lack of appetite, and weakness; often related to absorption of phosphorus in insecticides	Toothache and sore gums from oral infection or chronic periodontal disease
Garlic breath; often from industrial metal fumes such as selenium	Breath odors linked to chronic illnesses such as liver disease or diabetes

Likely to be Environmentally Induced	*Unlikely*
Metallic taste and blue line around border of gum, with sore gums and loosening of teeth; often indicates mercury poisoning	
	White patches on tongue; indicates thrush, a fungal infection

Skin

Chloracne (severe acne with pus-filled cysts) and blackheads; often from dioxin or PCBs	Dry, itchy eruptions filled with clear fluid, usually allergic eczema
Redness and irritation; often from acids, alkalis, copper salts, compounds of lead, silver, zinc, phosporous	
Change in pigmentation with scaling, white lines on fingernails, headache, and confusion; often from chronic arsenic poisoning	Yellow skin and eyes with accompanying dark urine and light feces from hepatitis
Bluish color from lack of hemoglobin in blood; often indicates aniline absorption from clothing dyes, paint removers, and inks or from phenols in disinfectants and preservatives	Body hives from insect sting or anaphylactic (severe allergic) reaction

Likely to be Environmentally Induced	*Unlikely*
Temporary flushing and tingling; often from monosodium glutamate or other additives	Full body flush with sudden weakness, difficulty breathing; may be allergic reaction to metabisulfites
Hives; often associated with chemicals such as formaldehyde and hydrocarbons in office-ventilation systems	
Cracks in skin on hand; often from solvents in duplication papers	

Heart and Respiratory System

Noticeable differences in heartbeat after heavy medication or prolonged use of home cleaning products	Irregular heartbeat after short-term exposure to medication or cleaning products; indicates hypersensitivity
Difficulty breathing; often during early stages of asbestosis, or from cotton dust, beryllium in electrical equipment	Cough, producing yellow or green sputum with fever, indicating infection
Coughing; often from inhalation of corrosive acids in bleaches and metal cleaners	
Wheezing; often from toluene in glues, inks, lacquers, and spot removers	

Likely to be Environmentally Induced	*Unlikely*

Gastrointestinal System

Stomach pains after eating with weakness, insomnia, visual impairment; often from working with rubber and other carbon disulfide products

Persistent abdominal pain radiating to back with nausea and vomiting; possibly pancreatitis

Nausea and vomiting with headache, tremors, pain in liver area; often linked to tetrachlorethane fumes in household cleaners and dry-cleaning plants

Genitourinary System

Genital rash that will not heal; sometimes from deodorants or contraceptive products

Painful genital rash that heals and recurs; possibly scabies (lice) or herpes

Painful or frequent urination; often indicating kidney dysfunction caused by chemical exposure

Burning urination with low back pain, chills, or fever; probably cystitis

Abnormal uterine bleeding; possibly from insecticides, herbicides

Discharge and itching from yeast infection such as candida

General

Edema (swelling) with lethargy, headaches, and nausea; often from pesticide exposure

Edema due to heart or kidney malfunction, or medication such as birth control pills

Likely to be Environmentally Induced	Unlikely
Edema with throat irritation; often from inhalation of aluminum in deodorants, antiseptics, or ammonia in cleaning products	
Numbness and trembling in extremities; often from methylmercury in fish	
Hyperactivity; often related to food dyes	
Inexplicable weight loss with irritability, constipation, insomnia, dizziness, headache; often from absorption or ingestion of lead	
Profuse sweating, thirst, nausea; often caused by phenols in anesthetics and antiseptics	Periodic sweating with mild nausea during menopause
Muscle spasms, impaired coordination, partial paralysis of arms or legs after breathing herbicide	

If you suffer any of these symptoms, your first action should be to see your doctor for a thorough physical exam, including blood and other laboratory tests. If the diagnosis is toxicity, or even if there is no definite diagnosis but your symptoms persist, our detoxing program will clear your system of poisons and diminish your symptoms. If necessary, we will also guide you through the various stages of physiological and emotional withdrawal.

THE DETOX ANSWER

Detoxification is a valid and effective answer to widespread toxic contamination—a major public-health problem of the twentieth century. The mid-1980s offer increased public awareness, a broad range of resources, specially trained personnel, and newly developed medical procedures to help combat the problem. Our program incorporates all these elements.

Interest in toxins first surfaced in the early 1800s, when Mattieu Orfila, a Spanish physician often called the "Father of Toxicology," attempted to classify the symptoms of various poisons. Orfila, who devised methods for detecting poisons in the body and contributed to the improvement of autopsy techniques, singled out this study as a distinct area of medicine and thus ushered in the era of modern toxicology.

World War I introduced chlorine and mustard gases, with their severe burning effects, particularly on the cornea of the eye. Public outcries started in 1924 against dangerous fumes in factories and against the dirt and rodent debris that contaminated wheat. The vast industrialization that followed World War II launched the era of synthetic chemicals. Substances as common as coal and carbon now appeared frequently in toxic forms such as carbon monoxide and formaldehyde.

Continuing "progress" led to the development of dichloro-diphenyl-trichloro-ethane (DDT), in 1944, and other potent insecticides, only to be followed by the discovery in 1946 that repeated administration of DDT caused irreversible brain damage in animals.

Reactions to all these new substances provoked growing concern about indoor and outdoor pollution and gave new importance to the hitherto minor field of science, toxicology. Unlike early toxicologists who were restricted to studying the effects of hemlock, opium, and snake venom, contemporary toxicologists must analyze old and new drug reactions and interactions, identify causative agents in illnesses, evaluate all forms of chemicals, and try to help the environmentalists re-create a safe and healthy world.

Because challenges appear faster than they can be met, researchers must often presume cause and effect of a toxin and an illness and suggest behavior based on that assumption. For example, there is no *absolute scientific proof* linking asbestos to the lung cancer known as mesothelioma; this form of cancer cannot be produced in laboratory tests with asbestos. There are no irrefutable data that ozone causes emphysema, that a lipstick ingredient can trigger depression, or even that heroin is addictive—yet there is enough empirical and circumstantial evidence for both doctors and the public to have little doubt that these statements are true. Illness and suffering cannot wait for 100 percent conclusive tests.

The Food and Drug Administration (FDA) which came into being in 1906 as part of the Department of Agriculture, was the first major government effort to assure that foods are safe and wholesome, that drugs are safe and effective, and that cosmetics are harmless. Such products are usually required to list their ingredients, be honestly and informatively labeled, and carry adequate warnings when necessary.

In 1970, the Environmental Protection Agency (EPA) was established to govern matters concerning air and waste pollution and to control the use of pesticides and all other substances affecting the environment. The Toxic Substances Control Act was added seven years later and requires manufacturers to report new chemicals and new uses of existing chemicals to the EPA, which then determines possible health risks and has the power to ban unacceptable substances.

Another government agency concerned with toxic substances is the National Institute for Occupational Safety and Health (NIOSH), which sets standards for industrial chemicals and eliminates on-the-job hazards. While experts often speak of "safe levels" of chemicals, radiation, and other pollutants, no one knows exactly what levels are dangerous because there is no absolute quantity of a toxin that is harmless to everyone at a set level. What is safe for one person might be lethal to another.

In light of the continuing influx of chemicals into our lives, toxicology might logically evolve into the science of detoxicology. Like its antecedent, detoxicology would study ways to treat toxic reactions and break old habits and addictions, but its main emphasis would be on ridding the body of unwelcome substances, cleaning up one's personal environment, reducing susceptibility, and preventing future toxic encounters.

The next chapter explains where toxins collect in the body, how they are handled by the prime organs of detoxification, and special precautions to take during pregnancy.

Chapter 3 describes the major groups of toxins—contactants, ingestants, and inhalants—and where they are commonly found.

Chapter 4 helps you to discover the source of your toxic ills and create a toxic-free environment in your home and at work. The solution to your symptoms may be as easy as eliminating certain foods from your diet or sleeping with closed windows to keep out polluted air. Whatever the situation, you will probably be approaching the problem from one of three perspectives.

1. You *do not know* the cause of your ill health so you want to explore all possibilities.

2. You *suspect* the culprit might be tap water, the herbicide your neighbor uses, or fumes from the new paint in the bedroom, and you want to confirm your suspicion.

3. You are *certain* you are reacting to a toxin, such as waste residues from a nearby dump site, but to escape the air, soil, and water contamination, you would have to change jobs, sell your home, and move away—and you want to find another alternative.

This alternative is presented in the General Detox Program for everyone concerned with diminishing toxic exposure. Choosing professional care is discussed, and the procedures in medical detox clinics are highlighted. The last section suggests safe substitutes to common toxic products.

In Chapter 5, addiction and withdrawal symptoms are explained, followed by nine specific detoxification programs for alcohol, caffeine, chemicals, cocaine, food, sugar, marijuana, medication, and nicotine. This section also presents case histories and detailed suggestions to ease your transition, physically and psychologically, from user to nonuser.

Chapter 6 is concerned with life after detox, takes a look at the latest scientific and technological innovations to ease your toxic burden, and outlines practical guidelines to maintain your newly found health.

2

The Body Toxic: A Biological Look at the Detox Process

Your body is a natural detox center, well equipped to do its job. The moment toxic substances enter your system, powerful forces go to work to neutralize and excrete them. Problems occur when too many toxins or too much exposure overloads the mechanisms and causes them to malfunction or shut down.

A combination of organs perform interrelated functions for the purpose of maintaining homeostasis—body balance and overall well-being. There are four primary systems responsible for the detox process: respiratory, comprising pharynx, larynx, trachea, bronchial tubes, and lungs; digestive, including esophagus, stomach, liver, gall bladder, large and small intestines, rectum, and anus; urinary, the kidneys, ureter, urinary bladder, and urethra; and dermal, the sweat and sebaceous (lubricating) glands of the skin. In addition, the placenta was once regarded as the fetus' personal detox organ, although it is now known that many toxins can penetrate this organ and cause harm.

How effectively these systems work to neutralize, transform, and excrete toxins depends on the individual's overall state of health and on the substances themselves. A new science called pharmacokinetics tracks the course of a given chemical through the body and determines how long it remains active, how hazardous it is, and how quickly and efficiently it is metabolized and excreted into urine from the kidneys, bile from the liver, sweat from the pores, or air from the lungs.

The more we learn about the routes of individual substances, as well as the general capacities of our various systems, the better prepared we will be to keep our detox mechanisms functioning at their best.

THE RESPIRATORY ROUTE

Maria G., a thirty-seven-year-old woman who lives in downtown Los Angeles, walks two miles to her advertising agency every morning. The polluted air she breathes is tainted with ozone—a photochemical oxidant, so called because it is formed by the action of sunlight on nitrogen dioxide and other gases.

She moves briskly, her eyes red and stinging from the smog, her breathing shallow and rapid as her bronchial tubes constrict to reduce intake of the harmful gases to her lungs.

Air first travels through the nasal tract, which is lined with a thick mucous membrane containing hairlike cells called cilia. These and other cells in the respiratory tract filter and absorb 50 percent or more of toxic particles that we inhale, making them unavailable to the body.

The remaining gases and insoluble particles are carried to the pharynx by wavelike action of the cilia, then proceed to the larynx, or "voice box," and on down to the trachea. The trachea, or windpipe, is a passageway to the bronchi, the two tubes that lead into the lungs. All along the route, particles are continually being absorbed into cells lining the walls.

Depending on the nature of the irritants, once they reach the lungs, they may be:

Ingested by phagocytes—cells that have the capacity to "eat" or engulf foreign particles and other cells.

Carried off in the bloodstream or lymph nodes to be circulated to other organs where they can cause damage.

Routed directly to the liver to be excreted in the bile.

Expelled in the exhalation of air.

Retained on the surface of the lungs where they can accumulate and cause illness.

Ozone is a deep lung irritant that in sufficient quantity can be harmful. In Maria's case, each time she breathes, the bronchial membranes produce more mucus to dissolve, dilute, and remove the particles. The increased mucus induces her to cough, a reaction which carries off some of the irritants.

If ozone exposure is habitual, the cells in Maria's lungs will continue producing excess mucus. Infections thrive in such a condition, lead to more mucus production, more infection, and a dangerous cycle. Repeated infections damage the delicate air sacs in the lungs and could bring on emphysema, chronic bronchitis, or possibly cancer.

Our bodies are so adaptable, however, that most of the air pollutants will be filtered and absorbed in her nasal cavities before they reach her lungs. Maria can aid the process by remembering to breathe through the nose, thus taking advantage of the protection provided by the upper respiratory tract. If Maria spends the rest of her day breathing relatively clean air and keeps herself in generally good health, her respiratory system should easily withstand the short exposure to airborne pollutants.

THE DIGESTIVE SYSTEM

George D. is a fifty-two-year-old stockbroker who takes a 5-milligram tablet of Valium four times a day to "calm his nerves." Known generically as diazepam, Valium belongs to a family of central nervous system depressants called benzodiazepines. Digestive fluids in the stomach and intestines break down drugs into simpler substances the body can handle, as part of the metabolic, or biotransformation, process.

Blood from the intestines carries the digested Valium to George's liver, his largest glandular organ. The liver weighs about three pounds and lies on the right side of the abdomen, beneath the diaphragm. As the body's first clearing house and main detox center, the liver performs more than 1,500 functions. Those specifically related to detoxification include the ability to:

Metabolize toxins, debris, and bacteria from the intestinal tract, destroying or changing them into harmless forms.

Produce urea, a waste product that removes poisonous ammonia from body fluids.

Filter harmful substances from the blood.

Metabolize and store vitamins and minerals.

Manufacture and secrete bile, a green, bitter fluid that is stored in the gall bladder, a small pouch near the liver. The bile dissolves fats like detergent cutting grease so that they can be diluted and digested. The bile also neutralizes acids.

Send metabolites—substances that have been broken down— into bile to be excreted in the feces.

Send metabolites back into bloodstream to the kidneys to be excreted in the urine.

While the enzymes in George's liver work to neutralize the tablet, some Valium metabolites escape to the brain. There they act on control centers to relieve muscle tension and anxiety. Until the metabolites are completely excreted, about twenty-four hours later, the Valium will do its work and George will experience a numbed sensitivity.

Since George has been taking Valium for only two months, has already begun to taper his dosage, and plans to be off the drug entirely in another two months, he should not suffer permanent damage. The liver is a remarkably resilient organ, and George can keep it healthy by maintaining a low-fat, balanced diet of fresh, additive-free foods. (See Chapter 4 for details of the detox diet.) With fewer chemicals to detoxify, the liver is available to absorb and store nutrients which are then made available to body cells.

If George were to continue taking Valium, the chemicals would gradually decrease the liver's ability to excrete bile. Toxins would not be eliminated as quickly or as efficiently, would stay in the system longer, and would do more damage. Some barbiturates, such as the sedative phenobarbital, can increase bile production and speed the elimination of toxins. These drugs are occasionally taken for this purpose but, because of their addictive nature, are not recommended.

One of the best ways George can protect his liver is to watch his liquor intake. Alcohol is perhaps the most common toxin to both attack and be metabolized by the liver. Heavy drinkers often develop hepatitis, an inflammation of the liver causing irregular bile flow. This condition may progress to cirrhosis, a fatty overgrowth of the tissues with accompanying degeneration of cells. When the liver ceases to function, so does the body.

Early danger signs are sensitivity to odors, darkened urine, nausea and vomiting, abdominal pain, absence of libido, weight loss, and fatigue. Skin discoloration and fever may follow.

The liver not only acts as the primary organ of detoxification, it is also a prime collection site for the accumulation of toxins, many of which are harmful. Carbon tetrachloride, used in fire extinguishers and some cleaning solvents, directly causes tissue deterioration and fatty infiltration. Phosphorus, a chemical in phosphorescent lighting, attacks veins near the entrance to the liver. Large doses of tetracycline, particularly in pregnant women, produce symptoms of hepatitis. Too much acetaminophen, found in Datril, Tylenol, and other painkillers, can result in severe damage. Repeated

exposure to the inhaled anesthetic halothane may cause hepatitis, and a small percentage of women develop chole-
stasis—stoppage of bile flow—from oral contraceptives.

Other toxins damaging to the liver include large amounts
of antidepressants, antipsychotic and anticancer drugs, ar-
senicals in fungicides and insecticides, and excessive amounts
of vitamins A, B3 (niacin), and B6 (pyridoxine). Aflatoxins,
poisons from molds found in contaminated grains and nuts,
particularly peanuts, can also be harmful.

If you are good to your liver, it will more than repay
you. Supplied with adequate nutrients and not overdosed
with chemicals, it will serve as your body's main detoxifying
agent for the rest of your life.

While the liver is the organ most active in the metabolism of toxins, the kidney plays the major role in eliminating
them. Let's follow George's Valium one step further.

THE URINARY TRACT

An hour or so after ingestion, liver-processed metabolites of
Valium travel through the bloodstream to George's kidneys.
These two bean-shaped organs are located in the back on
either side of the spine, just below the rib cage. Each is about
the size of a fist and weighs a quarter-pound. Their size is
no indication, however, of their importance. George's blood
composition depends heavily on how his kidneys handle the
nutrients, wastes, and toxins that pass through them. The
kidneys are hardy organs that process nearly 2,000 quarts
of blood a day, every day. They filter, cleanse, re-absorb
nutrients and water vital to the body, and excrete substances
the body cannot use.

The Valium metabolites now enter these two miniature
filtration factories, each consisting of more than a million
tiny, self-contained machines called nephrons. A nephron is
made up of bundles of blood vessels called glomeruli. Wavelike muscle contractions push the microscopic drops through
the glomeruli, extracting essential nutrients such as glucose,

minerals, and amino acids (compounds necessary to build proteins) and then returning them to the bloodstream through a tubule, a little tube.

The remaining material, including whatever is left of the Valium metabolites, drains into the ureter, a large tube which empties into the urinary bladder. The Valium metabolites are then expelled in George's urine. In this way the kidneys regulate the volume of body fluids and the exact composition of these fluids, preserving the body's normal chemical balance.

Several common drugs can be very damaging to the kidneys. One is aspirin in massive doses of, for example, forty tablets a day—twice as much as some arthritis sufferers take, yet even these people should watch for possible signs of kidney damage, such as changes in urinating habits, puffiness around the eyes, or pain in the small of the back.

Dr. William H. Bennett of the Oregon Health Sciences University in Portland warns that nonprescription painkillers such as Excedrin and Tylenol contain aspirin or acetaminophen, and used to contain phenacetin (now banned in over-the-counter medications). Heavily advertised as being safe, these ingredients can, with regular use, cause scarring and eventual kidney failure. Overuse increases the chances of contracting kidney infections and some forms of urinary tract cancer.

Large amounts of lead from environmental or occupational exposure, long-term use of certain antibiotics, and strong dyes frequently injected into the bloodstream for X-ray tests can also cause kidney damage. Doctors who prescribe these drugs or dyes weigh the benefits against the risks.

Like George's liver, his kidneys are remarkably designed. They can process a wide variety of environmental chemicals as part of their normal functioning and, if treated with care, will continue to be the body's main instruments of excretion. Thirst and a dry mouth usually indicate the kidneys require more liquid. Drink plenty of "clean" fluids, especially water, cut down on caffeine and other stimulants, and allow this vital detox center to perform its job.

THE SKIN

More than just the package in which our tissues live, our skin is also an essential detoxifying organ. If Maria and George were to exercise vigorously after exposure to their respective toxins, products of ozone and Valium would show up in their sweat.

The skin offers two routes of excretion. One is through the simple, tubular sweat glands that are found in most parts of the body. The second is via sebum, the oily skin lubricant secreted by the sebaceous glands. These grapelike masses surround each hair follicle and, like sweat glands, are stimulated to produce by a rise of body temperature.

Sweating is extremely good for you; it washes away dead layers of skin, opens up blood vessels and stimulates circulation, and supplies a major exit route for toxins. A 1982 study done by the Foundation for Advancements in Science and Education in Los Angeles found that as many toxins are excreted through the skin as through the urine.

Care of the skin should include washing regularly with mild soap and warm water, using lubricating oil or cream every day, using a minimum of cosmetics, protecting yourself from cold, windy weather, and avoiding sun exposure, particularly between 10 A.M. and 2 P.M., when the rays are most intense.

CROSSING THE PLACENTA

Our confidence in the efficacy of the body's detox systems has been well substantiated; and up until two decades ago, medical students were taught to regard the placenta, the organ that nourishes the fetus, as the fetus' personal detox organ—a protective barrier that let in needed nutrients but kept out harmful bacteria, allergens, and toxic chemicals. German measles and syphilis were the only known exceptions. The placenta, however, has proved to be less effective than we thought.

In 1962, the thalidomide explosion changed that view. The drug commonly prescribed as a "safe" tranquilizer was in fact a teratogen, a substance that damages the fetus. Babies of "thalidomide mothers" were often victims of phocomelia, or "seal limbs"—a name tragically descriptive of their malformed arms and legs.

Another drug, Bendectin, was used for twenty-seven years by more than 33 million women to curb nausea and vomiting during pregnancy. In June 1983, Merrell Dow Pharmaceuticals stopped selling Bendectin after a Washington, D.C., jury awarded $750,000 in damages to a Bendectin mother with a deformed baby. The company's lawsuit insurance soared, and a Merrell Dow spokesman announced: "We were forced for business reasons to take a safe and effective medication off the market."

It has not yet been proven in the laboratory whether Bendectin is safe or harmful. The FDA, whose review panel could find no association between Bendectin and birth defects, admitted, however, a "residual uncertainty." It would seem logical that if a drug had even a 0.1 percent chance of harming a fetus, a mother would be well-advised to find other ways to control her nausea.

Today, it is known that the placenta has mechanisms that can metabolize and prevent some toxic substances from reaching the fetus, though many pass through easily. University of Michigan Professor Arthur J. Vander, M.D., observes that, "Almost all environmental chemicals which can make it into the mother's bloodstream, will pass, to lesser or greater extent, into the fetus."

Once there, substances take various routes. The fetal blood brain barrier is immature, so some toxins, such as lead, tend to concentrate in the embryonic brain, where they may remain and cause brain damage. Pesticides and other chemicals often come to rest in fetal fat, where they may accumulate and cause later mental disorders or organ damage. Still other substances, such as alcohol, enter the fetal bloodstream, but cannot be metabolized by the embryo's underdeveloped liver. When the mother's blood alcohol level subsides, the alcohol passes back across the placenta to be excreted by the mother; how much remains in the fetus is not known, but the harmful effects of even small amounts of alcohol have been well proven.

During the last months of pregnancy, the fetal kidneys are working and excreting some substances into urine. The urine drains into the fluid surrounding the fetus and is expelled at birth. In general, however, the fetal detox mechanisms are incompletely formed, and the danger of toxic retention and damage is considerable.

The safest course during pregnancy is to avoid taking any nonessential medication, particularly in the first trimester—the three months following conception—when the fetus is most vulnerable to toxins. Even women thinking of becoming pregnant should avoid all unnecessary drug use. By the time they know they are pregnant, the damage may be done. The majority of drugs have not been adequately tested for possible teratogenic qualities, but this does not mean that all medication must be avoided. Pregnant women suffering such ailments as asthma, diabetes, epilepsy, tuberculosis, and heart disease may do more damage to the fetus by *not* taking prescribed medicines.

If you feel inclined to take any drug during pregnancy, check first with your doctor. The absence of negative findings, or a drug's current popularity, should not lull you into a sense of security about its effects. See Appendix B for a list of common drugs and their toxicity in pregnancy. Appendix C lists other probable teratogens: heavy metals, chemicals, vitamins, and substances related to personal habits, including alcohol, caffeine, and tobacco smoke.

Scientists also believe that there are serious consequences if a man takes drugs while he and his partner are trying to conceive or during intercourse if conception has occurred. The two theories are that drugs directly damage sperm, or, are carried in the semen, absorbed through the vaginal walls, and enter the mother's blood flowing to the fetus. Either may adversely affect the fetus.

Man or woman, if pregnancy is your goal, follow these five suggestions:

1. Stop or taper off all self-prescribed drugs used for the relief of minor or temporary symptoms. This is particularly important following a missed period or the instant a woman suspects pregnancy.

2. Interpret "drugs" to include: lotions and salves containing hormones, cortisone, or other compounds that may be absorbed through the skin; vaginal douches, suppositories, and jellies; rectal suppositories; medicated cough drops and syrup; and nose drops.

3. If female, tell your doctor as soon as you suspect you may be pregnant, and discuss any medicine, including vitamins and vitamin preparations, you are or will be taking. If you are seeing specialists, advise them as well.

4. Be sure to take any medication your doctor prescribes to treat serious or chronic problems.

5. Take the minimum effective dose if this can be determined. If not, take the exact prescribed dose and do not continue it longer or other than directed.

Research in this area is admittedly limited because the several factors contributing to our understanding of a drug's effect vary: when during gestation the exposure occurs; the amount that crosses the placenta; the route of administration; whether it tends to be metabolized, retained, or excreted; and perhaps most importantly, the genetic makeup of the unborn child. One fetus may react adversely to a certain drug while a second fetus will be immune.

Some people argue that animal tests are not reliable, but Dr. Joseph T. Morgan, an Oregon pediatrician and allergist, believes that studies done on animals can justifiably be extended to people.

"Humans inevitably prove to be *more* sensitive to teratogens than animals," he states. "For instance, in the case of thalidomide, humans are 60 times more sensitive than rats, 200 times more sensitive than dogs, and 700 times more sensitive than hamsters."

Pregnant or not, treat your body with the same care you would give to a fine-tuned, very expensive racing car; remember how hard it is to find replacement parts. Your reward will be in the way your detox systems disarm and dispatch the many toxic substances that invade your physical domain.

3

The Culprits: Three Major Groups of Toxins

The three major groups of toxins are categorized according to how these substances enter the body: Contactants are substances that touch and are absorbed into the skin; inhalants are airborne substances that we breathe; and ingestants are those we swallow. Less common are injectants, which enter the body by hypodermic needle, snake fang or insect sting, and suppositories.

The three major categories may overlap; an ingredient in cologne, for instance, could cause symptoms both by touch and by inhalation. A bug killer can be carried by the wind into our lungs, or consumed as a residue on foods. Cortisone can cause side effects whether swallowed, sprayed, injected, or rubbed on the skin. The purpose of this chapter is to describe toxins in the forms in which they are most frequently encountered and to help you locate "toxic hiding places" in your home or work environments. Chapter 4 outlines the appropriate steps to take in eliminating these toxins one by one. This chapter is descriptive; the next is prescriptive. Each is equally important; even minimal changes in routine or environment can be immediately beneficial.

CONTACTANTS

Our skin comes into contact with many toxic agents. Fortunately, human skin is not highly permeable and serves as an effective barrier separating us from most of our environment—but not all. PCBs, for instance, can penetrate the hands of electrical workers in less than half a second; protective rubber gloves only help for four or five minutes.

Toxins that enter the system through the skin can produce local reactions at the site of contact, systemic reactions elsewhere in the body, or both. For example, an ingredient in permanent hair dye could cause a scalp rash, nausea, and eventually affect the liver and kidneys. Carbon tetrachloride, a solvent found in some nail polish removers and insecticides might be absorbed through the skin, leaving it intact, and yet produce liver injury. People who heed initial warning signs, however, rarely suffer extensive damage from skin absorption.

The most common sign of a contact reaction is a rash; any sort of redness should be promptly investigated. It could also come from an ingestant, an inhalant, or an injectant, but the first step is to rule out the obvious.

The biggest clue toward discovering the culprit is location. An eruption on the face or neck might be from a perfume or makeup. A rash on the scalp or ears suggests a hair preparation. If your arms or hands break out, look for a cleaning agent, a substance you handle at work, even the rubber gloves you wear for protection. Another clue is the shape of the rash; sometimes it will appear in the perfect outline of a piece of jewelry, a bandage, or underwear.

Most contactants fall into the general categories of cosmetics, home and work supplies, and radiation.

Cosmetics and Toiletries

Americans spent $13 billion on cosmetics in 1983 and will spend an estimated $14 billion in 1984. According to the FDA, a cosmetic is any substance "rubbed, poured, sprin-kled, or sprayed on . . . the human body for cleansing, beau-tifying, promoting attractiveness or altering the appearance without affecting the body's structure or function." This in-cludes hair, nail, hand, body, scented, shaving, and dental preparations as well as makeup.

Almost any paste, powder, or lotion applied to the body can be absorbed through the skin and affect various tissues and organs. The FDA reviews each drug or medication before it is marketed but has no such requirements for cosmetics; hence, many of the preparations in common use contain one or a variety of troublesome agents. A single fragrance may have 100 ingredients; add this to a foundation makeup with 50 chemicals of its own, and your chances for a reaction greatly multiply.

Damage from cosmetics is usually minor and control-lable, yet often perplexing. Edith G., a musician, recently complained to Dr. Saifer of vaginal itching and irritation with no apparent cause. A careful examination, as well as allergy, blood, and other lab tests showed her to be in excellent health. Referral to a gynecologist ruled out other possibilities. The breakthrough came when Edith's twelve-year-old daughter developed a rectal itch that disappeared when the girl went to summer camp. It was finally traced to the home use of scented toilet tissue. As soon as the paper was dis-carded, both mother's and daughter's symptoms disappeared.

A more dramatic incident happened a little over a de-cade ago, when thirty-nine infants died from unknown causes, later linked to hexachlorophene, an antibacterial agent in certain soaps. Researchers discovered that this compound sticks to the skin, is absorbed into the bloodstream and at-tacks nerve cells, causing irreversible brain damage. The FDA has restricted its use to minimal amounts in prescription products only.

While the FDA does not test or pass approval on cos-metics, it does require ingredient listing. The burden then reverts to the consumer to read the label on all products and be alert for chemicals that might pose risks. (See Appendix D for a list of toxic ingredients in cosmetics.)

If, after reading the above, you still feel irresistibly drawn to the makeup or perfume counter, *always choose the cosmetic with the simplest and fewest ingredients* that you know you tolerate.

Home and Work Supplies

Individuals who use detergents and synthetic sponges to clean the bathroom sink, handle fresh newsprint, work with photographic materials, paints and solvents, or touch copier paper and other common office supplies are candidates for toxic skin irritation. Mechanical agents such as small particles of glass fiber, metal, or wood can get caught in the folds of the skin and cause swelling and itching. Even clothing, dry-cleaned or permanent-pressed with formaldehyde and other chemical finishes, can cause body dermatitis.

A newly widowed man, Elliot R., became Dr. Saifer's patient when his hands and fingers showed extensive blisters, scaling, and cracking. It differed slightly in appearance from the dry, itchy redness of allergic eczema. An extensive history revealed that Elliot had begun making scrapbooks of personal memorabilia and affixing them with glue containing the irritant benzene. He replaced it with nontoxic white glue and was able to continue pasting with healed hands. (Appendix E contains a list of common home and office products to avoid or use sparingly.)

The approach to any possibly toxic contactant is to stop or reduce usage as soon as you can. This may be difficult if it is a cosmetic or a cleaning product you depend on for your livelihood. Fortunately, there are safe substitutes for almost every household and cosmetic item (see Chapter 4), and in the case of occupational toxins, your employer is required by law to provide adequate protection.

The Occupational Safety and Health Administration (OSHA) reports that "two out of every five workers exposed to some form of skin irritant will develop industrial dermatitis—skin irritation or disease. Skin reactions range from slight reddening or mild itching to a rash or swollen, weeping, or open sores; however, nearly all of these dermatoses are preventable."

Most problems from contactants at home or at work can be solved by following these rules adapted from OSHA publication S-610, November 1981:

1. Immediately eliminate skin contact with irritating chemicals or substances.

2. Whenever possible, substitute products with low toxicity for more toxic substances.

3. Keep home and work areas clean, well swept, and dusted.

4. Keep yourself and your clothes clean. Clothing that has touched irritating chemicals or oils can cause dermatitis problems not only for the wearer, but for all family members, especially if the clothing is laundered at home.

5. Personal protective equipment, if used intelligently and kept uncontaminated, can minimize skin irritation. Safety goggles and face shields, natural or synthetic rubber gloves, sleeves, aprons, boots or shoes, and work clothing made from tightly woven fabrics can all help to minimize exposure. Cover the rims of safety goggles with strips of cotton fabric to protect against skin irritation and insure a tighter fit.

6. Wash hands and any exposed body parts with Ivory, pure castile, or plain, unscented soap immediately after exposure.

Radiation

As everyone knows, radiation hazards can result from contaminated food and water, but primarily affect the body through the skin. The simplest definition of radiation is that it is energy moving through space as invisible waves. Despite sensationalized stories of radioactive wastes and nuclear accidents, the average person has little to fear from moderate exposure.

Common sources of radiation include medical equipment, radioactive substances in building materials, nuclear and industrial technology, and cosmic rays from the sun. Rare sources are poorly sealed microwave ovens and minute emissions from computer terminals and color television sets.

The biggest contributors to radiation exposure, according to the FDA's Bureau of Radiological Health, are medical and dental X rays and the use of radioactive materials to diagnose and treat disease. While studies show that a large number of the 240 million X-ray examinations conducted annually may not be necessary, the bureau emphasizes that possible adverse effects of not having an X ray when it is needed can be worse than the radiation.

"Rad" is the unit for measuring a dose of radiation. A modern chest X ray, for instance, is produced with properly functioning equipment by less than one rad of energy. A single exposure to 100 or more rads, perhaps from a factory accident or malfunctioning hospital equipment, could cause radiation sickness. The first symptoms are nausea and vomiting, which subside, to be followed later by infections, hemorrhage, fever, diarrhea, loss of hair, and mental disorders from brain damage. Long-term results can be leukemia and other cancers.

"The average person not working around radiation equipment or living in a highly irradiated area is in no real danger of radiation poisoning," says radiologist Dr. Murry Schonfeld of Dallas. "All we're sure of is that the less radiation you get, the better off you are."

INGESTANTS

Any substance taken into the body by way of mouth and through the digestive system is an ingestant. This food-and-drink category includes *intentional* additives such as sugar and preservatives and *nonintentional* additives from pesticides, cooking utensils, and packaging materials.

Food

Carelessly handled food can become contaminated by viruses, bacteria, molds, parasites, animals, and insects. Our concern is, however, mainly with synthetic chemicals, which affect all commercial food to some degree. The fresh carrots sold at the supermarket are grown with chemical fertilizer, treated with insecticides, nematocides (to protect them from worms in the soil), and may have been chemically sprayed in the store to keep them fresh. Organically grown carrots bought from a reliable source are *less* chemically contaminated; they are not in any sense "pure." Even with the most meticulous care, they are still grown in soil, use air, and absorb water affected by technology. Minute residues, however, are not harmful to most people.

For purposes of rating chemical toxicity, there are three grades of food: processed, natural, and organic. Processed food includes junk food such as hot dogs, donuts, and candy bars, and TV dinners and canned goods that are low in nutrition and high in calories. Some have so many additives they are more artificial than real.

Natural foods have fewer chemical toxins than processed foods. They are grown with pesticides and chemical fertilizers but have no further additives. These include fresh meats, fish, fruits, vegetables, poultry, whole grains, nuts, cold pressed oils extracted without chemical solvents, and baked and cooked foods free of preservatives.

Organic products are supposedly grown without pesticides, chemical fertilizers, hormones, or antibiotics. No chemicals have been added at any time between the harvest and your table, but as stated before, nothing in our civilized world is pure. Organic foods are highly perishable and are sometimes contaminated by insects, dirt, fungi, and bacteria. The label "organically grown" is no guarantee that the product is free of pesticide or other residues, yet it is probably as close to nontoxic nutrition as we can get.

Unfortunately, there is a long list of foods that contain natural toxins, and in large quantities these can be dangerous. Some are fava beans; nutmeg; sassafras and licorice root; the seeds, pits, or leaves of cherries, peaches, apples, pears, and plums; and herbal teas such as chamomile, a potent allergen. (See Chapter 4 for further information about herbal teas.)

Intentional Additives   The FDA defines a food additive as "any substance that becomes part of a food product when added either directly or indirectly." Today, more than 3,000 chemicals are intentionally added to produce desired effects; 10,000 other compounds find their way into foods during processing and storage. It is estimated that the average American ingests one gallon of food additives yearly.

Additives tend to be present in much larger quantities than pesticides and therefore cause more general toxicity. Many on the FDA Generally Recognized as Safe list have not been adequately tested for harmful effects yet are widely used in the processing of food. (See Appendix F for food additives worth avoiding.)

Not all additives are harmful or toxic. When fresh food is unavailable, preservatives retard spoilage and ensure that we will not suffer hunger. Many commercial products are fortified with vitamins and minerals (often to replace those lost in processing), and "safe" or "natural" ingredients such as lecithin, an emulsifier, can sometimes improve texture and flavor.

A steady diet of these chemicals, however, is of questionable value. Occasional small exposures to any one additive should not cause problems, but the long-term effect of daily exposures to a variety of additives has not been laboratory tested. Some persons "miraculously" recover from "incurable" ailments such as arthritis or schizophrenia three or four *days* after they change to a diet of unprocessed foods.

Sugar Technically an additive, sugar is so widely used in foods and food processing—even cigarettes are laced with sugar—that it merits a section of its own.

One form of sugar, glucose, is essential to human metabolism. It is found mainly in fruits, flows through normal blood, and acts as a prime source of energy. There are more than 100 other chemicals called sugar, including those with familiar names like sucrose, maltose, lactose, dextrose, and fructose.

Sugars are derived from various sources. Cane sugar is a member of the grain family—wheat, corn, rye, oats, barley, rice, millet, and grasses—and is a common allergen. Persons sensitive to cane or corn sugar can usually tolerate beet and maple sugar, and sometimes honey, but few nutritionally oriented doctors would encourage eating large amounts of any sweet. Honey, maple syrup, cane and beet sugar, molasses (made from cane or beet sugar), corn syrup, fructose, and turbinado sugar all lack merit as nutrients and are, in fact, capable of harm.

Let us count the ways:

People who fill up on sugar foods have no appetite for balanced meals and may suffer malnutrition. Also, the lack of nutrients weakens the immune system, thus increasing susceptibility to viruses and infections.

Sugar can camouflage bacteria, chemicals, spoilage, and other toxins in food.

Sugar is a major contributor to tooth decay and gum disease, obesity, diabetes, hypertension and heart disease, hypoglycemia, vitamin deficiencies, and psychological disorders. It aggravates tendencies toward diverticulosis, colon cancer, and osteoporosis, and "feeds" fungi such as *Candida albicans* which multiply and cause skin, respiratory, bowel, and emotional problems.

Sugar can be addictive: it causes blood sugar levels to rise and supply quick energy, but when this soon passes the person feels droopy and in need of another "rush." Sugar in its many forms is not addictive the way narcotics are, but it can lead to both psychological and physiological dependency.

Binging on sugar causes a subsequent drop in blood sugar level that signals the brain to secrete adrenalin. This increases gastric activity which may cause indigestion and stimulate the formation of peptic ulcers.

Scientific evidence suggests that sugar can trigger migraine headaches.

The toxic effects of sugar have been underplayed, and reports to the contrary are misleading and dangerous. John Yudkin, M.D., Ph.D., emeritus professor of nutrition at London University, sizes up the situation well when he says, "If only a small fraction of what is known about the effects of sugar were revealed in relation to any other food additive, that material would promptly be banned."

Nonintentional Additives Toxins appear where you least expect them. One such place is your kitchen—even though you may carefully clean the counters with nontoxic cleansers, wash with pure soaps, protect your hands with gloves, and eat fresh, additive-free foods. It is beneficial to know the toxic potential of common kitchen utensils and products. (See Appendix G for a list.)

A woman recently came to Dr. Saifer's office with a self-diagnosed ailment: Eating food cooked or wrapped in aluminum products gave her diarrhea; yet even when she used iron pots, she still had minor bowel problems. As soon as

she learned that common table salt contains aluminum magnesium silicate (to prevent caking), she switched to sea salt,
which contains no added aluminum. Her symptoms
disappeared.

A general rule in shopping for kitchen utensils and products is not to look for bargains. The best quality merchan-
dise is usually the least likely to emit contaminants into
your food.

Other nonintentional additives include residues of sprays
and chemical fertilizer used on crops, particles from packaging materials, vapors or solvents from external sources,
lead solder or phenol in can linings, and colored dyes or
preservatives that leach into foods from paper cartons. The
General Detox Program in Chapter 4 suggests ways to cut
down exposure to nonintentional additives.

Drink

By now, consumers are well aware that pop sodas, diet
colas, canned fruit drinks, and other processed liquids all
contain dyes, salt, carcinogens such as saccharin, and more
chemicals than you can count or pronounce. The buyer must
weigh convenience, taste, and low-calorie advantages against
possible long-term toxicity.

More serious problems face us in the form of the three
ubiquitous beverages of modern society: coffee, tap water,
and alcohol.

Coffee; Caffeine and Other Toxic Additives
Caffeine belongs to a family of chemicals known as xanthines
(caffeine, theobromine, theophylline), which stimulate the
heart and central nervous system, act as mild diuretics, and
combined with other drugs, relieve pain. Heavy users of
caffeine report tolerance (more and more is needed to give
the desired effect), physical and mental dependencies, and
craving and withdrawal symptoms.

As with any controversy, there are articles, books, and

statistics to "prove" both sides. This is especially true with America's most popular drink, coffee; yet convincing evidence indicates that this liquid that goes directly into your digestive system can alter heart rhythms, raise blood pressure, aggravate a tendency toward lumpy breasts, affect mood and behavior, and possibly cause cardiac disease. Caffeine is found in lesser amounts in tea, cocoa, and certain soft drinks.

Caffeine is not the only toxin in coffee. Pesticides banned in the United States as too toxic to use are shipped to Third World countries where coffee is grown; trace amounts are returned to us in the beans. More than ninety-four different pesticides are used in Brazil, Colombia, Mexico, and Guatemala, where much of our coffee comes from. Once the coffee reaches the United States, only small samples are tested for illegal residues before marketing.

Another practice that has been questioned is the use of methylene chloride, a paint remover, to decaffeinate coffee. In her book *Cancer-Causing Agents*, Ruth Winter writes that "Methylene chloride belongs to a family of chemicals suspected, and in some cases, known to cause cancer." America's National Coffee Association argues that it is only carcinogenic if one drinks 12 million to 25 million cups a day.

Methylene chloride is a central nervous system depressant that the body metabolizes, but does not accumulate; yet studies on humans show that concentrated exposure to this solvent in the work atmosphere may cause loss of coordination and bring on headaches, insomnia, and heart palpitations. Tests conducted at Research Triangle Park in Durham, North Carolina, in 1982, found that high levels of methylene chloride cause liver cancer in rats. Probably for these reasons, the manufacturers of two major brands, Sanka and High Point, recently switched to ethyl acetate, a noncarcinogen, to use as a decaffeinating chemical.

There is a water method for extracting caffeine, but according to additive expert and author Beatrice Trum Hunter, "The process is less efficient than solvents, costs more, and makes the coffee more expensive. Hence, none of the major companies want to use it."

If you are a heavy imbiber, think about changing beverages or at least reducing your intake. Chapter 4 lists a variety of substitutes. They may not give you a caffeine lift, but neither will they cause symptoms and create a dependency. Chapter 5 will guide you safely and probably painlessly through caffeine withdrawal, and help you to maintain your goal of abstinence.

Water In random tests, the EPA sampled 500 public water supplies across the country and found that 20 percent—one in five—were contaminated with industrial chemicals. Manmade chemicals from manufacturing processes are dumped in rivers, buried in landfills, carried off in drainage pipes, and travel by wind, rain, and air currents to our drinking water.

So much has been written about the toxicity of tap water that many of us already boil or filter our home supply or drink our daily ration from glass bottles. These measures are prudent rather than paranoid. The Center for Disease Control reports that more than 4,000 cases of illnesses are linked to drinking water every year in the United States. (Appendix H lists common contaminants, and Chapter 4 suggests alternatives.)

Alcohol Used in moderation, wine, liquor, and liqueur can enhance the quality of life. A study that is frequently quoted, which appeared in the *Journal of the American Medical Association* in November of 1979, showed that people who have one to three drinks a day live longer than either alcoholics or abstainers. One theory is that alcohol reduces stress; another is that it protects against the buildup of fatty deposits on the inner walls of the coronary arteries and reduces the risk of heart attack.

Unfortunately, no reduced risk of coronary attack was found among those who consumed more than three drinks daily. English researcher Dr. David Horrobin cites American, British, and Swedish studies which show "if you choose people on the street at random, eliminate those with obvious alcohol problems, and screen the rest, up to one-fourth of these seemingly normal people will have . . . liver damage due to too much alcohol."

Not only are 33 percent of all accidents and crimes linked to alcohol, even so-called social drinking is fraught with personal hazards. The National Institute on Alcohol Abuse and Alcoholism warns that possible effects include greater risk of heart attack and stroke; increased danger of cancer of the digestive tract; birth defects; loss of memory and other brain functions; impaired liver functions; impotence; fatigue; symptoms from interaction with other drugs; and reactions to additives in alcohol.

Current medical research is slowly demystifying the physiological reasons behind alcoholism; meanwhile, more nutritional supplements and de-alcoholized beverages, support groups, and treatment facilities are available to help the person with problems. Chapter 4 suggests alternatives and Chapter 5 offers specific guidelines and resources if withdrawing from an alcohol dependency.

Medical Drugs A medical drug is a chemical compound used to aid in the prevention, diagnosis, and treatment of disease or other abnormal condition. This section discusses only those drugs taken in pill, powder, or liquid form. Cigarette smoke and other recreational drugs are included in the inhalant section.

Dr. F. Gilbert McMahon writes in the *Journal of the American Medical Association* in January 1983: "There is a misconception on the part of much of the public that medicines ought not produce any toxic or adverse effects. The truth is that any drug now available, whether over-the-counter or prescription . . . can produce serious side effects . . . [but] physicians prescribe drugs because they think the treatment will do more good than harm."

The point is well-taken. Patients who require medication for serious illness may have to weigh benefits against the possibility, or even the certainty, of side effects. As Dr. McMahon points out, "The gravity of an illness dictates the margin of allowable risks."

A wise doctor will not order you to take a drug or threaten to stop treating you if you decline. He merely explains the pros and cons and makes a recommendation. You must make your own decision if you are able to do so: Is it worth treating your body with chemotherapy to kill cancer cells? Are brittle bones from prolonged steroid use too high a price to pay for asthma relief? Is feeling numb and tranquilized preferable to being tense? In short—is toxicity worth the benefits? Often it is. No one can deny that medication in general does much more good for people than harm. The point is that whatever you decide to do, take the time to make an informed, balanced choice.

A list of every prescription drug with respective degrees of toxicity would be endless; results would fluctuate depending upon genetic makeup, age, medical history, and hundreds of other factors. Even knowing all that can be known about patient and drug, a doctor can never predict the outcome with certainty.

Here are five personal guidelines for using any prescription or nonprescription drug:

1. Do not take medication on the assumption that it "can't hurt and might help," or on the theory that if a little works, more will work better. All drugs carry risks.

2. Avoid time-release medications until you have established a minimum effective dose for yourself. You may be taking larger amounts than you need.

3. Remember that two active ingredients are not necessarily better than one. Your chances of toxic effects are greater with combination drugs.

4. Do not go on or off a prescribed medication without your doctor's approval.

5. Keep in mind that medication is a last—not a first—resort.

Some drug reactions are simply extensions of the desired effect. Too many barbiturates taken as a sedative, for example, can sedate you into a coma. Whatever drug you take, watch for signs of toxic blood levels such as nausea, dizziness, irritability, flushing and sweating, intense thirst, or a headache. Tinnitus, ringing in the ears, may be the first warning that you have exceeded your tolerance for even a "safe" drug such as aspirin. If you can identify the cause of your symptoms (a food or chemical reaction, stress, fumes, etcetera) you can avoid the trigger and no drugs may be needed.

Danger signs may also occur from interaction with food or other drugs, or when liver and kidney functions are impaired and the substance cannot be sufficiently metabolized or excreted. Cigarette smoke, a drug in itself, reduces the body's tolerance for other drugs. Caffeine is found in such common drugs as NoDoz, Vivarin, Excedrin, Midol, and Dexatrim, and may alter or intensify the effects of other medication.

Fortunately, most drug reactions are dose related, and the person who wants to get off sleeping pills, for instance, can detoxify himself by gradually reducing the dosage down to zero. When there is a physical or psychological dependency, the process is more complicated and may call for the withdrawal techniques described in Chapter 5.

INHALANTS

Today, "clean air" is a self-canceling phrase. Except in high mountain altitudes, remote communities far from all forms of urbanization, and some areas by the sea or ocean, the air we breathe both indoors and outdoors contains toxic substances. Susceptibility to their effects varies with each individual.

Indoor Air Pollutants

In late 1979, James L. Repace, an EPA official, carried a sensitive air-monitoring device as he went about his daily work, walked in smog, and drove through rush-hour traffic. To his surprise, the device showed the highest levels of air pollution while he was waiting for dinner in his own kitchen.

In the last few years indoor air pollution has been recognized as a growing and serious problem. Government researchers estimate that indoor pollution may be adding $15 billion to $100 billion annually to our national health-care costs. The reason, according to the 1982 California State Consumer Affairs publication, *Clean Your Room!,* is: "Building design changes intended to conserve energy, new materials used in construction, and the presence indoors of numerous hazardous substances are combining to make the indoor environment, where most Americans spend 90 percent of their time, an unhealthy place."

A ban passed by the U.S. Consumer Product Safety Commission in 1982 made it illegal to install urea-formaldehyde foam insulation (UFFI) in homes and schools but it has since been overturned and is now on appeal. That same year a New York science teacher named Michael Wagner won a $225,000 settlement from the company that installed UFFI in his home. Once sensitized to the chemical, Wagner had to make drastic changes in his habits. "I can't eat food cooked on natural gas," he told a reporter. "My wife can't use cosmetics, and oven cleaner is enough to make me sick for days. It's almost like living in a bubble—except the bubble can't be plastic because that contains formaldehyde." (See Appendix I for more about indoor pollutants, and Chapter 4 for ways to combat it.)

Outdoor Air Pollutants

A distinction is made between point and non-point sources of outdoor air pollutants. Point sources are those which can be identified as coming from a particular location or factory. They affect their immediate vicinity and can usually be controlled or regulated. A factory smokestack, a polluted river, and a chemical spill on the highway are all point-source emissions.

Non-point sources are more general and harder to locate and regulate. They include pesticides and fertilizers that drain off farmlands and vaporize into the air; emissions from stored wastes or landfills; and metal, rubber, and asbestos particles picked up everywhere by the wind. (Prime offenders are listed in Appendix J.)

The toxicology of air pollution is complex and widespread; virtually all Americans carry around traces of these materials, but most are not in large enough amounts to cause harm or symptoms.

Recreational Drugs

Nicotine, marijuana, and cocaine are the most pervasive of the "soft" drugs, but there is nothing soft about them. Surgeon General C. Everett Koop recently told *Time* magazine: "Cigarette smoking is . . . the chief preventable cause of death in our society and the most important public health issue of our time." Excluding "hard" narcotics such as opium, morphine, heroin, and methadone, which involve a physiological addiction, marijuana and cocaine are the two most widely used stimulants in the "underground" market.

Nicotine and Cigarettes   Nicotine, an oily, poisonous liquid found only in the tobacco plant, is pharmaceutically characterized as an organic nerve drug so powerful that a one-drop injection would cause instant death. The greatest danger of nicotine is its addictive nature, described in detail in Chapter 5.

More than 90 percent of cigarette smoke is composed of lethal gases, mainly carbon monoxide, hydrogen cyanide, and nitrogen oxides. These, along with some 4,600 other chemicals found in the tar—residue left in the filter after the smoke passes through—work very efficiently to cripple and ultimately destroy the human body.

Smoking is a major cause of cancer of the lungs, larynx, oral cavity, and esophagus. It can induce miscarriage, premature births, and birth defects, and heart, respiratory, and blood vessel disease. Studies also link smoking to cancer of the stomach and cervix and relate it to numerous other diseases.

Following a 1983 news story that low-tar and low-nicotine cigarettes are as toxic as other brands, the Chicago Tribune ran a cartoon of a patient protesting, "But I only like generic cigarettes." The doctor growled back, "How would you like generic cancer?" There is no such thing as a safe cigarette.

Marijuana Sometimes called the "love drug," marijuana comes from the *Cannabis sativa* plant, better known as hemp, and is the most popular illegal drug in the United States. Users claim it acts as a relaxant, releases inhibitions, increases self-confidence and creative abilities, heightens sensual awareness and self-acceptance, reduces aggressive tendencies and functions as an aphrodisiac. Quite an order for a weed.

Most scientists only recognize the sedative and euphoric effects of marijuana, although they do acknowledge possible medical benefits. Legislation now pending would enable doctors to prescribe the drug for use in alleviating the nausea associated with cancer chemotherapy and to arrest the inner eye pressures of glaucoma. Both uses have been proven in more than 100 responsible studies and are widely accepted.

Unfortunately, the drug also has less benign consequences. *Taking Care,* the newsletter for the nonprofit Center for Consumer Health Education, reports that marijuana smoke has as many cancer-causing substances as tobacco smoke. Chronic users get lung problems similar to those of cigarette smokers, along with personality changes such as chronic apathy, lack of motivation, and feelings of resentment and hostility.

The drug is also habit-forming and psychologically addictive, and may contain harmful molds, bacteria, or traces of paraquat, the herbicide that is lethal to humans if swallowed in doses as small as one-tenth of an ounce. In his book *Kids & Drugs*, Dr. Jason D. Baron reports that: "Outstanding scientists have published over 7,500 studies on the medical and psychological effects of marijuana that counter the widespread myth that marijuana is a harmless drug." (See Chapter 5 for a withdrawal program.)

Cocaine As a local anesthetic in medications such as Novocain or Marcaine, this drug is invaluable. More than 20 million Americans have used cocaine for nonmedical reasons, however, and some 5,000 new people try it every day.

Extracted from the leaves of the South American coca plant, cocaine is a central nervous system stimulant that produces a five- to twenty-minute euphoria, or "high," delusions, fantasies of power and importance, and a feeling of well-being followed by a period of depression. Some users soften the "crash" with alcohol, sedatives, or even heroin, all of which are extremely toxic combinations. "Speedballing," as the cocaine-heroin combination is known, can be fatal.

In small amounts, cocaine increases the heart rate and elevates blood pressure. Constant snorting damages the nasal membranes and causes nosebleeds, tender and irritated nostrils, along with anxiety, confusion, hostility, cold sweats, and insomnia.

Freebasing, a process of distilling pure cocaine from the usual street mixture of cocaine hydrochloride salt and inhaling or injecting it is extremely dangerous to the user. Because the cocaine is almost 100 percent pure, a single inhalation or injection could cause seizures, fainting, or convulsions. Some people experience toxic psychosis with hallucinations, paranoia, and feelings of extreme aggression. Continued use may lead to muscular paralysis, cardiovascular collapse, or respiratory failure—all of which can be fatal.

Regular users are as numerous as heroin addicts, but the addictions differ; cocaine withdrawal is characterized by more severe psychological and less severe physiological symptoms than heroin. Breaking the habit is difficult, but people are doing it. (See Chapter 5 for a withdrawal program.)

Eliminating or minimizing the use of toxins in your life can be a major undertaking, as in withdrawing from chronic cocaine use or alcoholism, or it can be as simple as substituting strong household cleaners for mild ones, or exchanging a shampoo with dyes and chemicals for one made of fresh eggs. Before you despair at having to part with a favorite substance, consider the possibility that it may be easier to do so than you think.

4

Help Your Body to Health: Beginning the Detox Program

You have now taken the first step in helping your body to health by thinking about some of your long-time habits and wondering if they merit the physical price tags. Perhaps you are already body-conscious; you take excellent care of yourself and have few bad habits, yet you still feel less than 100 percent. The General Detox Program that Dr. Saifer has developed is for every individual who breathes air, eats food, and drinks water. It has proven effective not only for chemically sensitive patients, but for those concerned with preventive health care and environmental cleanup. Taken in small steps, the seemingly grand-scale task of minimizing exposure to substances harmful to human biology will become routine and almost second nature.

We begin by recommending easy, basic ways to detoxify your home and work environments, then outlining a general program of nutrition and exercise to strengthen your toxic immunity. Procedures for detoxifying from alcohol, cigarettes, drugs, and other addictive toxins follow in the next chapter. If you need medical care, we help you choose a doctor or a care center. Finally, we discuss safe alternatives to common toxic household and personal items.

THE INDOOR CLEANUP

Those who suffer most from indoor pollution are infants, the elderly, the chronically ill, allergic persons, and expectant mothers, but we all pay a price whether it is obvious or not. Because the time lag between exposure and effect can range from instantaneous to months or even years, most of us fail to see a connection between indoor pollution and headaches, fatigue, skin rashes, mental disorders, and various undiagnosed symptoms.

One reason for this lag is that people react to substances and situations at some times and not at others, depending on the full-barrel concept described in Chapter 1. By reducing the total load of indoor toxins, you are much less likely to cross your tolerance threshold. Only when you overload your barrel do symptoms appear.

Carl T., a law student, consulted Dr. Saifer when he found himself confused, headachy, and unable to concentrate on his studies. A detailed history revealed that he had seen a spider in the shower a few days prior to the start of his problems and had begun spraying his bathroom with insecticide. He also smoked and wore strong-scented colognes to mask the tobacco odors—habits he had had for years with no noticeable symptoms. The pesticide was the last straw that pushed him over his tolerance threshold. When he stopped spraying, his symptoms diminished, but not until he stopped smoking and self-deodorizing as well did he regain the health he had had before he crossed his tolerance threshold.

Detoxifying your home and work environments is the starting point toward ultimately detoxifying yourself. To do this, you need to discover the source of your toxic ills. Study the world around you with an alert mind and acute senses. Note any sudden mood changes or physical symptoms related to the use of a particular food, smell, or substance. Determine if you feel better or worse in specific areas, even around certain people. Begin a room-by-room checklist to see if there might be a common denominator. Our general procedure follows.

At Home

Bedroom. Depending on the severity of your symptoms, your cleanup can be minor or extensive. Your eventual goal will be to replace synthetic furnishings made with materials such as polyester and foam rubber with natural fibers such as cotton, linen, and wool. Whenever you buy new bedding, linens, curtains, or solid furniture, it is important to remember that almost all synthetic fabrics outgas, or give off gases, and should be avoided if they have any "new" smell. Metal, glass, rattan, and hard wood furniture are generally nontoxic.

Whether or not you are hypersensitive, it is wise to keep garments that have been dry-cleaned, mothproofed, or washed in detergent in closed drawers and closets. Do the same with inks, cleaning agents, scented cosmetics, perfumes, and all products with odors. Store these items where they cannot pollute your air, or better yet, replace them with nontoxic substances. You may want to consult *Nontoxic & Natural: How to Avoid Dangerous Everyday Products and Buy or Make Healthy Ones* by Debra Lynn Dadd for lists of safe alternatives.

One retired couple baffled and frustrated Dr. Saifer when extensive histories revealed no reason why both husband and wife would get headaches and become quarrelsome every Thursday morning. It finally turned out that their Wednesday afternoon cleaning lady used polish containing aniline, a toxic dye, to shine the husband's three pairs of shoes which sat by their bed. Inhaling the fumes all night affected their central nervous systems, and they awoke with symptoms. When the polishing stopped, so did their irritability and headaches.

If your home has forced air heating, gas and oils from

the furnace can be toxic. Close the vent, seal with tape, and use a portable electric heater. You can insulate your bedroom from gas and other fumes in the rest of the house by putting felt weather-stripping around the door.

Consider not allowing people into your bedroom who smoke, wear strong perfumes, or carry the fumes of their profession. A painter might reek of solvents; a surgeon might carry antiseptic odors on his hands and clothes.

Kitchen. Substitute toxic cleaning products with the toxin-free alternatives recommended in the last section of this chapter and, if possible, replace a gas stove with an electric one. All stoves should be vented to prevent buildup of the air pollutants associated with burning fuels. Minimize use of synthetic furnishings, such as Dacron curtains, which outgas considerably in the heat generated by cooking. Look instead for cotton, linen, or unbleached muslin. Formica and most hard plastics do not outgas and are safe. The rule we always come back to is: If a product has an odor, avoid it. (See Appendices F and G for food additives and utensils which may cause problems.)

Bathroom. Dispose of all drugs with overdue expiration dates and those you no longer use. Keep toiletries, cosmetics, and cleaning products inside drawers or cabinets. Use a charcoal filter on the shower to cut down chlorine fumes. In fact, consider using a charcoal filter on your entire water supply to eliminate chlorine and other toxic substances.

General. A home loaded with knickknacks is a home loaded with dust, and dust not only causes allergies, it may also contain fragments of lead, asbestos fibers, particles from tobacco smoke, and other toxins. Remove as many extraneous objects as you can bear to part with, keep your home clean and well vacuumed, and open windows when the air is clear. If you use a wood-burning stove or a fireplace, be sure the flue, vent, or chimney is unobstructed and that the smoke goes outside. Combine common sense with your newly gained knowledge to create a toxin-free residence.

At Work

Initiating your detox program is somewhat more difficult on the job, where you may not always have a say. In general, however, you should be able to reduce contamination without causing too much inconvenience to your co-workers.

In an Office. Modern energy-sealed buildings recirculate air that carries fumes from glues, copier solvents, fluorescent lighting, newsprint, personal cosmetics, tobacco smoke, air fresheners, adhesives in building materials, paints, fire-retardant chemicals on furnishings, and more. (See Appendix K for list of toxins related to particular occupations.)

To help reduce exposure:

1. If you work in a new building, try to secure a desk by a window that opens (many do not), or in a corner away from machines and gas heating outlets.

2. If your office is enclosed and poorly ventilated, and fresh air is unavailable through windows, shut the doors or screen off the area, if possible, and keep an air purifier going twenty-four hours a day. To test ventilation, simply light a match. If the smoke moves away, the air is moving. If the smoke just hangs there, the pollutants are hanging around, too.

3. In any building, try for an office near the top; the higher the floor, the cleaner the surrounding air.

4. Do not smoke, and keep a sign on your desk requesting others not to smoke around you. If they refuse, talk to your employer. You have a legal right to a healthful working environment.

5. Whenever possible, choose nontoxic furnishings free of foam rubber, plastic, and synthetic fibers. If you have a chair of plastic or foam rubber, cover it with heavy construction-grade aluminum foil. Hard surface floors, not carpeted,

help minimize pollution. Get rid of dust-catching bric-a-brac and books that are never used. Keep your desk free of papers, which gather dust and may emit ink or copier fumes.

6. Use pens, staples, and paper clips, which are odorless, instead of glue. Minimize use of odorous correction fluids and cements and try to give freshly printed or copied papers a chance to dry before reading them.

7. Request that pesticides and strong cleaning solvents not be used in your area. If this is impossible, consider doing your own cleaning with acceptable products.

8. Have pure water available to drink.

9. If you use a video display terminal (computer), try to sit at least twenty inches from the monitor. The face of the screen carries a charge of positive ions that draw negative ions from the surrounding air and may make you feel sluggish. Sun-Flex Co. Inc., 20 Pimentel Court, Novato, CA 94947, sells a volt-free, antiglare protective shield that covers your screen, cuts down glare, and eliminates static electricity. A single shield costs $65, with discounts for groups.

Be sure your chair has adequate back support. Take a fifteen-minute fresh air break every two hours, or every hour if you are working under pressure. The roof of your building, unless a chimney is discharging smoke, will have cleaner air than the street below. Another suggestion is to alternate two hours at the computer with two hours of a different kind of work.

10. Stretch or exercise your limbs as often as you can, breathing deeply.

At a Plant Around Chemicals. Your employer is required by law to ensure that you are trained to do your job safely and are given whatever protective gear you need. Respirators, portable breathing masks, are a basic defense against poor ventilation and stagnant or toxic air. To obtain a current list of government-approved respirators, write to Testing and Certification Laboratory, NIOSH, 944 Chestnut Ridge Road, Morgantown, WV 26505.

Check with your supervisor to be sure that a chemist or lab technician is collecting and evaluating the various dusts and gases around you. The air should be continuously monitored and not allowed to reach toxic levels.

Get a copy of the Material Safety Data Sheet (MSDS) for any substance that you work with. This is a document, filled out by the manufacturer, that supplies information— hazardous ingredients, health risks, spill or leak procedures, and protection measures—for each chemical.

If your employer does not have a particular MSDS on hand, he must make a written inquiry to the manufacturer within seven working days of your request. The documents, OSHA Form 20, are supplied by the federal government and are available from any OSHA office or branch of the U.S. Department of Labor.

At a Plant Around Machinery. Ventilation is extremely important, as welding, soldering, cutting, mining, and blasting all generate dust or fumes. Each process that releases dust should have a local exhaust along with the general ventilation system. Individual respirators, safety shields and goggles, rubber aprons, gloves, and earmuffs may all be necessary.

THE OUTDOOR CLEANUP

The toxicity of outside air is obviously much more difficult to control. During summer, heat makes chemicals more volatile, and such compounds as car exhaust and tar in roofs and roads add to the level of air pollution. Natural plant chemicals called terpenes, or resins, also saturate the air and afflict the sensitive.

Heat stress can overload the toxic barrel and cause

symptoms, particularly when a person is working with pesticides, herbicides, preservatives, and chemical fertilizers in agricultural areas or inhaling the potent petrochemicals of road building and construction work. Complete body covering, including face mask, gloves, and boots should be worn around any type of spray or smoke fumes. Frequent rest periods in cooler temperatures, cold beverages, and salt pills may help.

Cold weather may be a similar trigger, reducing the body's defenses against airborne toxins. For outdoor work in chilly climates, thermal clothing and convenient areas of warmth should be provided. Write to the NIOSH information system for futher information on job or industrial hazards: NIOSHTIC, Robert A. Taft Laboratories, 4676 Columbia Parkway, Cincinnati, OH 45226.

There are several other ways the average person can minimize exposure to outdoor pollution:

Keep your house windows closed on smoggy days; do not go out unless you have to.

If you walk or jog, stay away from heavy traffic areas; better to exercise indoors than to inhale pollution. Early morning air is usually the freshest.

Drive with car windows closed in heavy urban or industrial areas and open windows where the air is clear. Purchase your next car with a recirculating air conditioner and heater so that outside odors and air pollution are diminished. You may want to use a charcoal air filter which can be plugged into the cigarette lighter.

Drive in times of minimal traffic and take less traveled routes. Steer clear of construction zones and newly sprayed farmlands. Heavy vehicles such as mobile homes, trucks, and buses give off the most fumes, so avoid trailing them. Keep at least half a city block between you and the car in front.

Consider joining a political action group lobbying for stricter air quality control. (See environmental resources in Appendix L.)

If you can, schedule your vacations by the seashore, on the ocean, or high in the mountains where the air is fresh and clean.

THE GENERAL DETOX PROGRAM

The Detox Program we present is designed to alleviate the physical and emotional symptoms of toxic exposure. Now that you have freed your home and work environments of toxins, it is time to free yourself—through nutrition, exercise, and relaxation.

Step One: Nutrition—Build Up Your Toxic Resistance

Vast research collected by Massachusetts University Professor Edward Calabrese and science writer Michael Dorsey shows that one of our best defenses against potentially harmful chemicals is in our diet. In their book *Healthy Living in an Unhealthy World,* Calabrese and Dorsey state, "Nutritional deficiencies increase one's susceptibilities to the toxic and carcinogenic chemicals around us. So, an easy way to build up one's resistance to the environment is to eat a balanced diet." Further evidence shows that not only do certain foods resist toxins, they also aid in metabolizing and excreting them.

To build up your resistance, increase your intake of the following food groups. Bear in mind any food sensitivities or allergies you may have.

Fruits, Vegetables, and Whole Grains. These provide nutrients and enzymes essential to the body's detoxifying mechanisms and also hasten passage of stool. When stool rests for long periods in the body, its chemical residues can be damaging. Increase your vegetable fiber intake gradually, however, or you may suffer gas, bloating, and diarrhea.

Carotene, in carrots and leafy green vegetables, converts into vitamin A in the human intestine and appears to neutralize certain carcinogens; vitamins C and E, in grains and green vegetables, help the body deactivate and get rid of the cadmium in cigarette smoke, as well as the pesticide DDT. Vitamin E also strengthens the lining of the respiratory tract, which makes it a more efficient filter of air pollutants such as ozone and nitrogen dioxide.

Grains, especially wheat products, contribute selenium to our diet, an element that enhances the ozone-fighting qualities of vitamin E, and helps protect cells from the destructive effects of polluted air. There is enough selenium in bread, garlic, pasta, cereals, vegetables, chicken, and seafood so that selenium supplements are not needed.

Other Carbohydrate Foods. The chief function of carbohydrates is to supply energy, as well as to aid the body's digestion and metabolism processes. Foods such as brown rice and whole grain breads and cereals release tryptophan, an amino acid that has a relaxing effect on the brain. According to researchers at Temple University in Philadelphia, tryptophan also serves as antidepressant and painkiller and helps reduce cravings.

Bananas, turkey, and natural cheeses are other good sources of tryptophan; it should only be taken as a food, however, never in pill or capsule form. Too little is known about possible harmful effects.

Fish and Poultry, Lean Meat, Low-Fat Dairy Products. These are protein building blocks which raise levels of adrenal hormones that help you fight the blues, which are a common component of withdrawal syndromes. Protein also strengthens hair follicles, sometimes loosened by chemical exposure. Eggs are a good source of lecithin, which aids in the assimilation of fat, helps relieve anxiety, and induces sleep, but they may be a cholesterol hazard. Large amounts of dairy foods are a potential hazard to the allergic or lactase deficient person, so eat adequate but not excessive amounts of protein.

Your liquid intake should include:

Milk. Another source of tryptophan and also of calcium, milk helps protect the body against carcinogens in heavy metal dusts. (Caution: milk is the most common food allergen.)

Vegetable Juices, Fresh Fruit Juices, and Clean Water. Raw juices such as fresh carrot and celery provide vitamins, minerals, and enzymes that are quickly assimilated into the bloodstream, where they are available to the body for all metabolic processes. A liquid intake of at least six glasses a day is essential for kidney excretion.

Herbal Teas. These can be very soothing to nerves that have been overstimulated by drugs, but some are potentially harmful. Check the list of teas to avoid in the section on alternatives at the end of this chapter.

Along with increasing your intake of the above, *reduce* consumption of foods which lower your resistance or add to your toxic load:

Red Meats. They not only contain hormones and additives, but carcinogens accumulate in their fats. Cancer specialist Dr. Virginia Livingston-Wheeler states that, "Most flesh foods are highly contaminated . . . with cancer-causing microbes and dangerous chemicals. Cattle are fed with chicken feces because the latter is rich in protein. The sick cattle are the ones that are sent to be slaughtered first. When we eat the cells of infected animals, their constituents become a part of us."

Excess animal protein can slow down digestive processes and lead to general sluggishness and apathy, as well as food fermentation in the intestinal tract, which causes formation of ammonia, a suspected carcinogen. Large amounts of protein also provide too much nitrogen, which puts undue stress on the kidneys and may slow down excretion processes.

Organ Meats. Hormones used to fatten cattle and pesticides from grains tend to accumulate in high amounts in organ meats, especially liver.

Canned Foods. Cans often contaminate their contents with lead from the solder in their seams or from phenols in the lining. The food in them usually supplies empty calories—bulk with minimal nutrients—and may contain dyes, preservatives, flavor enhancers, and other dubious additives.

Salt. Sodium causes bloating, water retention, and slows down the excretion of urine. The only time salt may be necessary in your diet is after strenuous exercise in hot weather.

Sugar. Sugar and other refined foods such as white flour take four times as long to pass through the bowels as raw fruits and vegetables; this slowed excretion has been linked with diverticulosis (little cracks in the intestinal lining) and colon cancer. Small amounts of sugar impede white blood cell functioning and lower the body's resistance to toxins; large amounts can paralyze white blood cells.

Saturated Fats. Fats in the form of butter, some cheeses, and heavily marbled meat accumulate in tissue and make handy depots for chemicals which might otherwise be excreted. The chemicals retain their toxicity and enter the bloodstream as the fats break up. When fats are present in the stomach, they slow down food passage into the intestines, resulting in greater absorption of toxic additives.

General. Refined and processed foods are usually oversalted, oversugared, high in additives and preservatives, and nutritionally bankrupt. They increase the toxic workload and deprive the body of nutrients needed to maintain its metabolization and excretion systems.

When food addictions are a problem, lessen cravings by eating five or six small meals a day instead of two or three big ones.

The Detox Diet   This regimen is not designed for those who use any drug, including alcohol, caffeine, nicotine, or medication. You cannot rid your body of chemicals while continually adding new ones. This diet restricts your eating for a minimum of four days—and after that for as long as you choose—in order to flush out toxins. A clean-out of chemicals obviously requires the intake of chemical-free foods. The Detox Diet works under the following principles:

1. Your intake of nutrients must be balanced.

2. Use only pure, simple foods with no additives and preferably "organic," that is, grown without pesticides or chemical fertilizers. Oils should be cold pressed, extracted without chemical solvents.

3. Use no coffee, vinegar, alcohol, sugars, processed or premixed foods. Herbal teas without additives are permitted.

4. All food must be cooked in spring or filtered water in pots of stainless steel, glass, porcelain, or cast iron. Avoid aluminum, nonstick, and plastic cookware.

5. Foods must be rotated. The traditional rotation diet means eating food from a different food family for each of four days, and then repeating the cycle; for instance, on Monday you might have corn, crab, and citrus fruits; on Tuesday you could have apples, fish, and beans; four days later, on Friday, you would repeat the food groups you had on Monday; on Saturday, repeat the foods you had on Tuesday, and so on. The Detox Diet is a more flexible version of the rotation diet, because you rotate foods once every two days instead of four.

A complete list of food families and the Four-Day Rotation Diet can be found in *The Type 1/Type 2 Allergy Relief Program;* here is a partial list to help you get started. Eat foods from families marked with an asterisk on the first and third days, and from food groups without an asterisk on alternate days. Foods in parentheses are to be avoided.

Plant Families

Apple—apple, pear, quince

*Banana—banana, plantain

Buckwheat—buckwheat, rhubarb, sorrel

Cashew—cashew, mango, pistachio

*Citrus—grapefruit, kumquat, lemon, lime, orange, tangerine

*Composite—artichoke, chicory, endive, lettuce; safflower, sesame seed, sunflower seed and their oils

Goose Foot—beet, (beet sugar); chard, spinach

*Gourd—cucumber, melon: cantaloupe, casaba, honeydew, watermelon, pumpkin, squash

*Grain—barley, corn, millet, oats, rice, (sugar from cane), wheat, wild rice

Grape—grape, raisin (wine and wine vinegar)

*Laurel—avocado, bay leaves, cinnamon

Legume—alfalfa, kidney, lima, navy, pinto, soy, and string beans, carob, garbanzo, lentil, pea, peanut

Lily—asparagus, chives, garlic, leek, onion

Morning Glory—sweet potato, yam

Mustard—broccoli, brussels sprouts, cabbage, cauliflower, mustard, turnip

Myrtle—allspice, clove, guava, paprika, pimiento

Olive—olive and olive oil

*Palm—coconut, date, hearts of palm

Papaya—carrot, celery, dill, fennel, papaya, parsley, parsnip

Pepper—black, white pepper

Pineapple—pineapple

*Plum—almond, apricot, cherry, nectarine, peach, plum

*Potato—eggplant, chili, red and green peppers, potato, tomato

*Rose—blackberry, boysenberry, raspberry, strawberry

Animal Families

Amphibians—frog

Birds—chicken, duck, eggs, partridge, turkey

*Crustaceans—crab, lobster, shrimp

Fish—any without shells

*Mammals—beef, butter, cheese, gelatin, milk, veal, lamb, pork, sheep

Mollusks—abalone, clam, mussel, oyster, scallop

When you begin your diet, keep in mind that you will be getting less variety and therefore should eat larger portions than usual. This is not a weight-loss program, but as your body starts to rid itself of pesticides and other toxins stored in fat, you will most likely shed pounds. The menus that follow are four sample days of a Detox Diet. Season everything lightly, use sea salt instead of table salt, and as little as possible at that. Go to each meal hungry. You will discover tastes you never knew existed! Recipes for those dishes marked with an asterisk follow the sample menus.

Day 1

Breakfast: Hot oatmeal with roasted pecans
Fresh peach slices or frozen peaches without syrup
Fresh orange juice

Lunch: Avocado stuffed with shrimp on bed of lettuce

with safflower oil and fresh lemon juice as
dressing
Cucumber sticks
Melon in season

Dinner: Roast lamb without skin
Baked potato
Fresh artichoke hearts filled with pureed
zucchini

Day 2

Breakfast: Hot whole buckwheat with unsweetened
pineapple chunks
Half papaya

Lunch: Tuna with alfalfa sprouts, celery, carrots, asparagus, green onions, beets
Filberts

Dinner: Broiled skinless chicken with grapes
Steamed string beans with sliced, toasted
almonds
Cabbage salad with raisins and sprinkle of ol-
ive oil
Baked apple with walnuts

Day 3

Breakfast: Half grapefruit with shredded coconut
Hot wheat cereal with sliced bananas
Lemon balm tea

Lunch: Tomato-beef broth*
Grilled natural ricotta cheese on rye crackers
Berries in season

Dinner: Roast veal with wild rice
Eggplant lightly fried in sesame seeds and
corn oil
Chilled hearts of palm

Day 4

Breakfast: Fresh pear halves
            Scrambled eggs with chives, cooked in soy oil
Lunch: Steamed broccoli florets with beet strips
            Sliced skinless turkey breast with green sauce*
            Fresh pineapple
Dinner: Parsley-chicken soup*
            Fresh fillet of sole with onions sauteed in ol-
              ive oil
            Fresh peas
            Puree of sweet potato and carrots
            Waldorf salad*

If you follow the principles outlined here, you will feel lighter, bouncier, and more alert within forty-eight hours. After four to seven days of the Detox Diet, you can begin experimenting with and expanding the variety of foods and combining ingredients. Discover new recipes in health-oriented magazines and cookbooks on our Recommended Reading List.

RECIPES

Tomato-Beef Broth

1½ pounds very lean beef brisket, trimmed
 4 large tomatoes, quartered
 1 quart water
    Sea salt to taste

Cut beef in five parts for faster cooking. Add tomatoes, water, and sea salt, and simmer two to three hours, or until meat is tender. Strain and refrigerate the broth. Skim off any remaining fat. Save beef for eating cold, or when rotation permits, with Green Sauce.

Green Sauce

    2 large celery stalks plus leaves of 3 stalks
    ¼ cup parsley
    1 cup green onion pieces
    1 cup olive oil
    Sea salt to taste

Chop all greens together or in food processor. Add oil and salt, stir. Delicious with cold chicken, turkey, or boiled beef.

Parsley-Chicken Soup

    2 tablespoons olive oil
    1 large onion, chopped
    1 large carrot, chopped
    1 celery stalk, chopped
    1 whole chicken (no liver)
    1 tablespoon parsley, finely chopped and packed
    1 quart water
    Sea salt and pepper to taste

Saute onion in oil, add all other ingredients and simmer, covered, for an hour or until chicken is tender. Cool, remove chicken, refrigerate, and skim off fat. Dice ½ cup chicken and add to soup.

Waldorf Salad

    2 cups diced, skinned, sweet green apples
    1 cup chopped celery
    ½ cup chopped walnuts
    ½ cup seedless raisins
    2 or 3 teaspoons apple juice

Mix apple juice with raisins and walnuts (or in food processor for about three seconds) to make a nutty, meaty dressing. Mix all ingredients. No mayonnaise needed. Serve immediately.

Supplement Your Diet with vitamins and minerals. The vital role of vitamins in protecting you against toxicity cannot be overemphasized. As soon as you finish the active Detox Diet (which allows no vitamins or minerals because of their coloring and additives), begin a program of supplementation. Be sure to check the ingredients list and buy vitamins and minerals that have as few fillers and binders as possible. Check with your doctor, however, before taking any vitamin or mineral supplements; a little is healthy, a lot may be harmful. Those with heart disease or high blood pressure should be especially cautious.

Dr. Saifer recommends the following daily dosages:

1. 500–1000 milligrams of vitamin C (ascorbic acid)

The *Journal of Nutrition*, June 1978, indicates that high levels of vitamin C reduce the toxic effects of chlorinated pesticides. Guinea pigs that received 800 milligrams of ascorbic acid broke down the rat-killer dieldrin twice as fast as guinea pigs who were given only 100 milligrams. Researchers believe vitamin C stimulates the liver's production of metabolizing enzymes and speeds detoxification of chemicals. Evidence also shows that: vitamin C reduces the formation of kidney lesions and modifies the toxicity of vinyl chloride; respiratory filters impregnated with 20 percent ascorbic acid give workers more protection from heavy metal dust; and smokers have diminished vitamin C levels in urine, suggesting that their vitamin C intake should be increased to compensate for extra needs.

A recent Russian study showed that vitamin C aids in eliminating excess fluoride from the bodies of industrial workers.

2. 200–400 IU (international units) of vitamin E

Animal studies reported in *The New England Journal of Medicine*, May 1983, indicate that the effects of ozone exposure from air pollution, office copy machines, and so on, can be minimized by oral supplementation of 200 IU of vitamin E daily. Both C and E act as antioxidants, protecting body cells from the damage caused by air pollution and radiation.

Vitamin E may also reduce red blood cell susceptibility to chemical and lead toxicity and may help protect the liver from harmful chemicals.

3. 50–100 milligrams of vitamin B complex

The B vitamins, nature's tranquilizers, reduce the effects of stress and calm the nervous system. They also help enzymes do their job of speeding body processes, and thus increase the efficiency of the detox systems. Vitamin B_6 (pyridoxine) is especially important; it acts as a natural diuretic and stimulates body production of a chemical known as 5-hydroxytryptamine, which seems to help prevent the depression that accompanies drug withdrawal. Large amounts of B_6—1,000 milligrams or more—can be toxic.

4. Optional

Take a multiple vitamin-mineral supplement containing zinc and calcium, as these minerals help reduce the effects of air pollution on lung tissue.

Step Two: Exercise—Detox While You Sweat

When you begin the Detox Diet, develop a regular exercise routine and spend no less than thirty minutes a day doing whatever activity you choose. Swimming in clean, not heavily chlorinated water, is one of the best detoxifying exercises. It uses all muscles without emphasizing any one area, burns calories at the rate of 300 to 1,000 an hour, and causes you to sweat and exhale toxins at a rapid rate. Jumping rope, running, dancing, rowing, playing tennis, and all active exercises that work up a sweat are recommended. Weightlifting and isometrics may develop your muscles but will not particularly stimulate your detox systems.

The detoxifying benefits of exercise are numerous:

Heavier breathing stimulates the mucosal blanket that coats the respiratory tract and filters out toxins.

Exercise increases heart rate and circulation, speeds the flow of air in and out of the lungs, and hastens toxic excretion in sweat, bile, stool, and urine.

Fifteen or more minutes of a vigorous workout stimulate the adrenal glands to produce epinephrine (adrenalin) and norepinephrine (noradrenalin)—hormones that speed up blood circulation and all metabolic and excretion processes.

Intense exertion mobilizes fat in which toxic chemicals such as DDT and other pesticides are stored, making them available for excretion.

Hearty activity also stimulates the pituitary gland to release endorphins—morphinelike substances that relieve pain and induce a sense of well-being. Endorphins are said to cause the euphoria known as "runner's high."

In all stages of detoxification, periods of fatigue, depression, or temptation may induce you to revert to old habits. Exercise will combat all of these states.

Tips:
Begin your daily workout with gentle reaching and stretching movements.

Wear warm clothes to stimulate circulation and increase sweating.

Practice inhaling deeply while you exercise. Diaphragmatic breathing actually works as a vacuum, pulling toxins out of the body and excreting them in the exhaled air.

Shower to freshen skin and rinse away "toxins."

Step Three: Relaxation—Hang Loose

The same hormones secreted during exercise—epinephrine and norepinephrine—also come from stress. Small amounts are beneficial as mood elevators and arousal enhancers, but in quantity these hormones cause our barrels to overflow and produce toxic symptoms.

In his book *Nutrition, Stress, and Toxic Chemicals*, Dr. Arthur J. Vander states that adrenal hormones released during times of stress increase salt retention and slow down kidney excretion. They reduce the body's resistance to carcinogens and stimulate the liver to release sugar into the blood, causing conditions that diminish the efficacy of the body's detox mechanisms.

There are a number of techniques that work well to balance out the ravages of stress: massage, relaxation tapes, biofeedback, yoga, self-hypnosis, water therapy, meditation, and listening to music. *Well Body, Well Earth,* the Sierra Club environmental source health book, contains an excellent section on visualization, another form of relaxant therapy.

Spend at least twenty minutes a day slowing down your brain and body processes. When you relax, the body's self-healing mechanisms go to work to replace cells, tissue, and whatever else toxic substances have damaged.

Researchers at the University of Kansas compared physiological relaxation responses of two groups of people. One group had been practicing meditation for four to ten years; the other relaxed without meditating. The result: No difference in heart rates or blood pressure. Conclusion: You get the same relaxing benefits from setting aside twenty uninterrupted minutes to sit quietly as you do from meditation.

You have now taken the three most important steps toward starting the detox process. Keep in mind that these initial actions are changing and correcting lifetime habits and patterns; it can take up to two years to show maximum benefits. That should be your time frame.

If you are discouraged or frustrated at not achieving faster results, you may want to supplement the program by talking with a family counselor, a support group, a psychotherapist, or a doctor. Detox resources are steadily increasing in source and variety: hospitals, private clinics, counseling centers, community outreach programs, state and local medical clinics, self-help support groups, employee assistance programs, religious institutions, individual therapists and physicians, and more. (See Appendix L for a resource list.) Don't be afraid to seek medical help to determine how to proceed from here.

IF YOU OPT FOR PROFESSIONAL HELP

Whether you feel you can easily break your habit or anticipate serious withdrawal symptoms, your first act should be to consult a family doctor who knows you, your family, and work situation, and belongs to an active medical network. He will also know which patients have been helped by which techniques, doctors, and institutions, and which have not. He is aware of his colleagues' reputations, and should be knowledgeable about community resources or be able to refer you to someone who is.

Before you commit to any group or institutionalized treatment, visit the facility and interview someone on the staff, preferably a doctor, to be sure you share the same goals. Take along a relative or close friend, if you can. Two sets of eyes and ears are better than one, and your friend may later recall details you missed or offer a fresh perspective.

Here are the basic questions to ask, whether you are considering an inpatient (residential) or outpatient (live at home) program and whether you have chemical dependencies or simply want a detox regimen.

1. What is the daily routine? Does it include:
   - A physical fitness program?
   - Family education and counseling?
   - Review of life-style?
   - Discussion of goals?
   - Nutritional guidance?
   - Educational lectures and films?
   - Instructions in stress reduction and relaxation techniques?

2. Are there library and recreational facilities available? (A desirable program should include several or all of the features listed in questions 1 and 2.)

3. How many hours, days, or weeks is the program? (This should be ascertained or at least estimated in advance, so that you know the time and costs involved.)

4. Who is in charge and what is this person's background? (The director should have a scientific background and be able to guarantee at least an initial evaluation, ongoing supervision, and a final consultation with a medical doctor.)

5. Who will I see? How often? (The therapist need not be a physician, but should have had sufficient training. Handling human emotions at this time may call for an extremely sensitive approach.)

6. How many M.D.s are on the staff? Registered nurses? (A high percentage of doctors and nurses on staff usually means better medical care.)

7. How many hours of psychotherapy will I receive? Group or private? (Psychotherapy needs vary with the individual. For people with problem dependencies, three hours a week of either group or private therapy is minimum.) Led by whom? What are the leader's qualifications? (Again, adequate training is essential.)

8. What lab tests will be done and what will the results show or not show?

9. How is detoxification accomplished? (All steps of the detox process, including the use of medication, vitamins, or any other therapy, should be explained in detail.)

10. What kind of drugs or medication do they use? What are the possible side effects? (The fewer drugs the better, but they should be available for persons with severe problems.)

11. Do they advise vitamins or other supplements? (Be sure these are medically approved for you by your doctor.)

12. Are medical services available at all times for acute problems? (Chemically dependent persons should ensure that twenty-four-hour help is available.)

13. Is there a twenty-four-hour crisis line for outpatients? (This, too, should be available, or at least someone should be on call.)

14. What are the fees? Should I expect extra costs? Is a deposit required? Is any refund offered? In what circumstances? Will I get all this in writing? (Fees vary, but expect to pay $230 to $350 a day for inpatient treatment, not including extras, and $80 to $100 a day for outpatient care. Everything should be spelled out on paper.)

15. Is the program covered by medical insurance? MediCare? (Most medical insurance will cover at least part of the detox procedure but this is determined individually. As of January 1984 all Blue Cross and Blue Shield plans provide alcohol- and drug-abuse benefits as part of the standard package offered to national accounts. MediCare covers some institutions and not others.)

16. Is aftercare available? What does it consist of? How long does it last? Does it include family members? Is there a charge for it? (The best programs provide a year of aftercare, which includes regular "alumni" meetings with support groups. Family members are usually encouraged to attend. Follow-up care may also offer a specific number of medical consultations and overnight stays at the facility. There should be no cost or very minimal charges.)

What to Expect in a Detox Unit

Linda G., age 39, had worked at the perfume counter of a large department store for eight years. She did not smoke, but drank coffee and martinis, and had started taking Seconal (a barbiturate) to help her sleep. Lately, she had been feeling stressed, tired, and lacking energy. Everything she did seemed to be a chore.

A friend suggested that the constant inhalation of perfumes, along with the intake of caffeine, alcohol, and sleeping pills, processed foods, car exhaust fumes, air pollution, and other chemicals in the environment, might be slowly poisoning her system and sapping her energy. Linda took a friend's advice to try a detox program.

She consulted her family doctor, a traditionalist who did not think much of the idea but who finally sent her to a toxicologist who served on the board of three detox centers—two private and the third part of a hospital. The toxicologist explained to Linda that all three programs were for chemically dependent people who wanted to break away from cigarettes, alcohol, coffee, medical and recreational drugs, as well as for people like herself who simply wanted to clear their bodies of foreign chemicals. The difference was that the hospital program provided *continuous medical care* during withdrawal, while the private, nonresidential centers did not. They offered *part-time care* during withdrawal, but mainly continued the detox process after withdrawal or treated patients who had physical complaints but were not necessarily addicted.

Since Linda did not need hospital care, the toxicologist sent her to interview the directors of the two different nonresidential programs. Here is what she learned.

Clinic A. This program costs $600 for a month of twice-weekly visits, which include eight colonic irrigation treatments. They are not mandatory but strongly encouraged. The colonic machine consists of a water tank with two tubes and carries the water (through a speculum inserted into the rectum) much deeper into the body's plumbing system than the standard enema.

The patient strips from the waist down, is covered by a

sheet, and lies on his back while the trained therapist flushes large quantities of warm water, thirty gallons or more, into the colon. The water is drained repeatedly over the course of an hour or so and is often accompanied by external massage of the colon.

Some therapists prefer cold water because it contracts the colon, enabling it to expel toxic gases quickly. It may be too much of a shock, however, for patients with heart problems. The treatment should not hurt, but the patient may feel chills, nausea, weakness, and light-headedness, and will of course experience the discomfort of diarrhea. At best, the patient will feel invigorated and relieved of anxiety, headaches, or other minor symptoms. His skin will feel flushed and tingly, his eyes may be bright.

No one there told Linda, however, that colonic irrigations are highly controversial. One study, published in 1981 in the *New England Journal of Medicine*, reported that the design of the colonic machine makes fecal contamination unavoidable. Syndicated talk-show host Dr. Dean Edell reported in March 1984 that there have been deaths from infections spread by improperly cleaned colonic equipment, that there are no regulations or uniform standards governing colonics, and that in his opinion, "Positive results are due to the placebo effect."

According to San Francisco gastroenterologist (stomach, intestines, and liver specialist) Dr. Jeffrey Aron, colonics can also "stretch the muscles of the colon and actually break the muscle fibers so that the colon eventually loses tone."

Anyone considering this type of therapy should first consult a physician or request information from the American Colon Hygiene Association, 7770 East Camelback Road, Suite 23, Scottsdale, AZ 85251.

Other treatments at Clinic A include once a week chiropractic adjustments, acupressure and acupuncture sessions, and a specific diet that includes daily intake of at least two quarts of water, juices, and herbal teas to flush out the urinary system. A natural cathartic of psyllium seed and seaweed powder—a "bulkifier"—is taken to carry out wastes from the colon.

The exercise regimen is designed to increase the elimination of toxins from the lungs, sweat, and sebaceous glands, and instructions are given for epsom salt baths followed by skin brushing with a loofah sponge to remove dead cells and draw toxins out of the skin. The colonics, acupuncture, acupressure, and chiropractic adjustments are done at the clinic; everything else happens in the patient's home.

Clinic B. This facility has quite a different program. It costs $1,350 to $2,000 for two to three weeks of treatment, two to five hours a day. The patient first gets a physical examination by a medical doctor, plus hair analysis, blood, neurological, and psychological tests. The clinic also offers lab procedures for analyzing human tissue, fat, and skin secretions and determining the type and amount of chemicals present. These tests run from $75 to $150 and are covered under most major medical policies. No drugs or medication are used during this period, unless specifically approved.

All treatments take place on the premises and include beginning-to-aerobic exercise classes, followed by forced sweating in a sauna kept at around 140 degrees—50 degrees lower than most health club saunas— and ventilated. Water, salt, and potassium are supplied to avert salt depletion and dehydration.

The patient is responsible for getting balanced meals and plenty of sleep at home, while the clinic dispenses detoxifying vitamin and mineral supplements, plus two to eight tablespoons of polyunsaturated oil which somehow nudges the body into releasing toxins by exchanging toxic fat for "clean" fat. The patient also takes gradually increasing doses of niacin (vitamin B3), from 50 up to 5,000 milligrams a day, to aid the breaking down of fat tissues. The use of niacin is generally followed by flushes which look like sunburn, but diminish in a few days. (Again, Linda was not warned that repeated use of 2,000 milligrams or more of niacin could cause liver damage.) Detoxification is effected through breath exhalation, full body sweating, and urinary and fecal excretion.

Underwhelmed at the thought of colonics, Linda chose Clinic B. As directed, she brought along an inventory of all prescription, over-the-counter, and herbal medicines she had been taking for the last year. She started the program tired, depressed, and feeling guilty about the expenditure of money. On the sixth day, she had a flashback: "I felt spaced out, dizzy, floating off into nowhere. I began to see things—faces, people dancing, people smiling and screaming at me. I got a strong taste of licorice—it reminded me of the nitrous oxide gas the dentist had given me—and also of strong, black coffee. I could feel and taste the toxins pouring out of me."

She felt disoriented for several days following the episode, then began to notice the cloudiness disappearing and her energy coming back. "I had another flashback," she recalled. "I was enveloped in the sickly smells of perfume and rotting vegetation. I tasted coffee again, but it was more of a chemical taste. My abdomen ached, and my Caesarean scar felt red and inflamed. It was sore and tender for several days. Then I started to feel good—really good, almost as if I were back in my teens."

After twenty-two days, Linda passed her physical exam and was amazed to find that she had no trouble concentrating, and could give much more cohesive answers to the psychological tests. All her senses—hearing, taste, sight, touch, and smell—were keener, and her skin seemed to glow.

The authors of this book are not endorsing any commercial program or detox center, nor have we investigated them for honesty, safety, or efficacy. In Linda's case, however, she did feel that the program had returned her to health, and she was determined to change her life-style in order to maintain it.

For everyone, this means continuing the detox regimen presented in this chapter: a clean diet, an active exercise program, and avoidance, as much as possible, of toxins in air, foods, products, and water. Eliminating such substances is easier when there are simple and effective substitutes. The alternatives we suggest will insure a safer, toxic-free life-style that supports your newly regained health.

ALTERNATIVES TO TOXIC PRODUCTS

Your final action in the General Detox Program is to substitute nontoxic substances for common products and to accustom yourself to using them. This often requires forming new habits as you begin your new life-style or start your withdrawal. Familiar situations that make you want to reach for a pill, a drink, a smoke, a cup of coffee, or even a hot dog are easier to manage with a handy alternative. Let's start in the home.

Cleaning Products

Americans have become obsessed with cleanliness. We are constantly reminded that we cannot possibly live the good life unless our bathrooms and kitchens are free of bacteria, our furniture mirror-shiny, our windows crystalline, and our floors scrubbed. These results can only be achieved, we are told, with commercial products containing strange-sounding "miracle" ingredients that are never explained.

As a nation, we spend $5 billion annually on obliterating smudges, streaks, scratches, and spots. We mask natural smells with synthetic deodorizers and replace dirt with chemical contamination. If the garbage is not lemon-fresh, nor the toilet water blue and reeking of pine, we instinctively reach for the aerosol.

In a word, we overuse manufactured cleaners. We need to cut down on sprinkling, spraying, and chemically sanitizing both our homes and ourselves. All complex commercial products should be replaced with unscented soap, baking soda (sodium bicarbonate), vinegar, washing soda (sal soda), and borax. Wear gloves while working and open windows.

To get you started, here are some specific alternatives for common consumer goods.

Air deodorizers. Fill dishes with kitchen herbs, spices such as cloves and cinnamon, potpourri, or baking soda, and distribute about room.

Carpet shampoos. Remove small spots with club soda or a mix that is half white vinegar, half cold water. Sprinkle baking soda, then vacuum. Test a tiny area for colorfastness before applying to a larger area.

Disinfectants. Clean area with soap and hot water. When someone is sick, boil objects handled by patient. Use unscented white facial tissues and disposable eating utensils. Combine zephiran chloride, an antiseptic available at drug stores, with equal parts of borax dissolved in water to protect surfaces against molds.

Drain cleaners. Flush drain with hot water followed by a heaping tablespoon of baking soda or half a cup of vinegar. Follow with several cups of cold water. Open blockages with a small plumber's snake or other mechanical device.

Fire extinguishers. Keep a commercial fire extinguisher on hand for major fires and use as a one-time emergency procedure; in minor kitchen fires, pour on baking soda to suffocate flames.

Furniture polish. Vacuum furniture or dust surfaces with slightly dampened cloth. Apply olive or lemon oil onto wood and leather. Sprinkle lightly with cornstarch and rub to a high gloss.

Glass cleaner. Add a tablespoon of white vinegar to three cups of water. Spray or rub on surface and wipe dry.

Laundry soap. Grate pure bar soap, such as Ivory, add water, and liquefy in blender. Store in a tight glass container.

Pesticides. Better cleaning habits, especially of hard-to-reach areas under counters and behind cabinets, can reduce food supplies for pests. Remove water sources by repairing pipe leaks and clogged drains and clear away clutter such as old rags and newspapers, which provide shelter. Seal cracks in walls and screens and clean damp areas around the yard that may be breeding grounds for mosquitoes. Use mechanical methods, such as mouse traps and fly swatters.

Don't waste your money on ultrasonic pest eliminators. These small electronic units claim to "bombard your home with silent sound that turns roaches and other pests into nervous wrecks and makes them vacate the premises." *Consumer Reports* of June 1983 found that pests "appear to condition themselves to the sounds quickly. The devices have no observable effect." Some better possibilities exist: for ant repellant, sprinkle red pepper in populated areas and plant mint by front and back door of house; for beetles, place a bay leaf in flour, rice, cookies, or any container with wheat products; for fly killer, make fly paper by boiling equal parts of sugar, corn syrup, and water together and spreading mixture on brown paper, and keep garbage covered and windows screened; for rodents, bait small traps with cheese, bacon, peanut butter, or chocolate candy, locate points of entry and plug them with steel wool, and keep cats fed well so they will be strong enough to catch rodents (do not expect them to work harder because they are hungry); and for roach killer, keep kitchen meticulously clean, especially the floor, and sprinkle boric acid in crevices and corners (cucumber rind will not kill roaches but will repel them).

Scouring powders. Dampen area to be cleaned, sprinkle with baking soda. Scrub with sponge. For difficult jobs, use borax first, then baking soda. Also, a solution of half vinegar, half water, will clean most surfaces.

Clothing

Unless you are unusually sensitive, you need not discard your polyester and synthetic fabric clothing, but when you buy new garments, look for natural fibers. These are cotton, linen, silk, and wool. Earth colors will have fewer dyes. All fabrics are treated with starch or gelatin sizing, which usually can be neutralized by laundering the garment before wearing it the first time. Cut down on permanent press, water repellant, flame retardant, and Scotch-Guarded fabrics, as their finishes generally contain formaldehyde and other chemicals that outgas.

Commercial mothproofing is dangerous to your health, as well as to the moth's. Scare away the pests by packing a lavender sachet in the drawer with your sweaters and other woolens. Do not put away soiled garments, particularly those with food stains, as these are invitations to dinner. Before folding or storing clothes, brush them as firmly as the fabric will allow to kill larvae and eggs.

Polish leather garments and shoes with a light vegetable oil and dry with a flannel cloth.

Beauty

The main problem with cosmetics and toiletries is that we tend to find something we like and use it repeatedly, causing toxicity, building up a sensitivity, or simply irritating the skin. Scents, in particular, can be offenders. As with food and drink, it is important to rotate cosmetics—using the same product no more than two or three times a week—and give the body a chance to excrete toxins before they accumulate.

"Hypoallergenic" or "natural" products often contain chemicals that not everyone can tolerate. Do not assume that any product is universally "safe." You may even want to make your own. A few suggestions follow:

After-shave. Use diluted lemon juice, or make a mixture of 1 cup strong mint tea and 1 tablespoon vodka.

Bath or body powder. Cornstarch, tapioca, or arrowroot are effective as bath powders.

Deodorant. Mix baking soda with a pinch of cornstarch to

increase absorbency, or dunk a cotton ball in cider vinegar and rub on appropriate areas. Once the vinegar dries, its smell disappears.

Eyebrow pencil. Charcoal from a wood fire works for some people. Be sure that the stick is free of ashes and cinders and that the eye is closed when you apply it.

Eye drops. Lie down, place cucumber slices or wet tea bags on closed eyes, leave for five to ten minutes.

Face cream or moisturizer. Use a sparse coating of olive, coconut, or any light oil.

Facial mask. Apply yogurt, raw papaya, egg white, or a paste of oatmeal and water. Let dry, splash on cool water, and blot dry.

Hair conditioner. Beat one egg yolk with half a cup of plain yogurt. Rub mixture into freshly washed hair, let stand three to five minutes, then rinse thoroughly with clear water.

Hairsetting lotion. Mix half a cup of warm water with three teaspoons fine sugar.

Mouthwash. Gargle with cool mint tea or a teaspoon of baking soda dissolved in a glass of water.

Perfume or cologne. Dip a cotton ball in peppermint or lemongrass oil and apply lightly.

Shampoo. Beat one egg into one cup lukewarm water. If you desire suds, add grated soap or liquid made from pure soap. For a rinse, mix one tablespoon vinegar or lemon juice with one pint of water.

Shaving cream. Coat skin with soap or a light vegetable oil, or use an electric razor.

Skin cleanser. Grind almonds, oatmeal, or sunflower seeds in blender. Pour a teaspoonful into the palm of your hand and add enough water to make a paste. Pat gently over face twice a day and rinse with cool water.

Toothpaste. Use baking soda or mix half a cup of baking soda with a teaspoon of salt. Whip in blender to a fine powder. If you are on a salt-free diet, brush with plain water; it is the act of flossing and brushing, the mechanical dislodging of food debris, not the toothpaste, that does the most good.

Eating

You have already begun your nutritional overhaul and have started to replace canned and processed foods with fresh foods. If pesticides concern you, wash fruits and vegetables with biodegradable detergent or soak them for five minutes in a mixture of a fourth of a cup of vinegar and a gallon of water, then rinse and dry thoroughly. Peeling helps, especially for highly sprayed fruits such as apples and pears. While the combination of peeling and cooking may lose nutrients, it does get rid of some of the pesticide residues.

Growing your own food is more trouble than it is worth if you live in a large city or industrial area. The air and water will be contaminated and will infiltrate your crops. If your air, water, and soil are reasonably clean, however, and you have the time and energy to care for a garden, you will surely enjoy the taste, freshness, and relative purity of homegrown produce.

Along with rotating your foods, as discussed in the Detox Diet, satisfactory substitutes are available for the two most popular additives.

When it comes to salt, train your taste buds to require less. Sea salt has fewer added chemicals than table salt. Many delicious herb seasonings are now on the market. If you buy canned or frozen food, try the low-salt or sodium-free varieties.

As for sugar, raw honey and pure maple syrup are less chemically contaminated than white sugar but otherwise offer no advantages. Reports have come in linking aspartame (NutraSweet) with headaches, dizziness, blurred vision, seizures, menstrual cramps, depression, and fatigue. A New Jersey researcher, Dr. Howard E. Warne, has found evidence that aspartame can cause hyperkinesia—abnormally increased muscle movement—in children.

Large doses of saccharin have been linked to bladder cancer in animals, and although there is no conclusive proof that the sweetener causes cancer in people, it seems a good product to avoid.

The best and safest sweetening alternatives to refined white sugar are fresh fruits, such as pureed bananas, dried fruits (especially dates), and unsweetened fruit juices. Top cereals with a sauce of apples cooked in apple juice, or dried fruits soaked in a bit of water and pureed in the blender. Learn to cook with sweet spices and herbs such as cardamom, coriander, ginger, cinnamon, nutmeg, basil, and mace. A sprinkle of grated coconut or raisins will make food taste sweeter.

Drinking

Replace soft drinks with half fruit juice, half soda water. Unless you know your water to be clean, drinking from the tap is not always advisable. Carbon filters cost from $150 to $500, and a recent EPA study found that the best models fit under the sink and hook into your faucet. Filters selling for less than $150 may improve taste and smell, but generally do little to lessen chemical contamination.

Other possibilities are:

Boiled tap water. Boil uncovered for ten minutes and allow water to cool, until no more steam comes off. This removes some of the chlorine and volatile pesticides.

Distilled water. This will be free of all chemicals, including essential trace minerals. Fine for short-term use, distilled water is not desirable as a long-term substitute.

Spring water. This will be chlorine-free, but may contain

excess minerals and pesticide residues. Still, it is the best alternative available. Buy water in glass bottles, as plastics leach chemicals into the contents.

(Coffee and alcohol are discussed in a later section on recreational drugs.)

Medication

Many medical drugs are indispensable and lifesaving. Others are virtually unnecessary and may be replaced by changing habits, reducing exposure to toxins, eliminating mental and physical stresses, improving diet, instituting an exercise regime, or using herbal or nonchemical substances. Some medical drug dependencies can be cured by cutting down frequency or dosage, or simply stopping usage.

The suggestions that follow are not intended as medical advice, but as a guide in working with your doctor. Orthodox physicians may look askance at so-called healers, but the choice is up to you and you should know the alternatives, namely, herbalistic, holistic, homeopathic, naturopathic, and orthomolecular medicine.

Herbalistic. Herbal practitioners are generally not M.D.s, but are skilled at preparing remedies from the seed, fruit, flower, leaf, stem, root, bark, or wood of various herbs. Contemporary herbalists offer diagnosis, counseling, and specific herb mixtures for common and complex ailments; their recipes combine experience and intuition. Advocates claim to get lasting benefits with no drug side effects.

Both herbalists and physicians warn against self-prescribing; many persons end up in hospitals suffering the toxic effects of homemade potions.

Holistic. This widely and frequently overused term should not be ignored because of its popularization. The Association for Holistic Health in San Diego, California, states that "maintaining good health involves much more than just taking care of all the various components that make up the physical body." Rather than concentrate on one specific symptom, organ, or body system, holistic medicine treats the whole person—mind, body, soul—and how that person interacts with his environment.

Holistic health practitioners are either M.D.s or D.O.s (doctors of osteopathy) who focus on disease prevention through stress reduction, developing proper eating and living habits, and a positive emotional outlook.

Homeopathic. The principle of homeopathy, a branch of medicine popular at the turn of the century and currently being revived, is that "like cures like." In other words, a substance that produces symptoms of a disease in a healthy person will cure a person who has the disease. The bark of the cinchona tree, for instance, produces malarialike symptoms in healthy people but also alleviates the symptoms of malaria victims. The reason, according to homeopaths, is that minute doses of the substance tend to activate and strengthen the body's own healing mechanisms.

Homeopathic doctors are mainly M.D.s who derive their remedies from animal, herb, and mineral sources, which have potent but nontoxic effects. The homeopathic practitioner only administers one pure medication at a time, never a mixture of two or more substances that might act together to cause an unpredictable response.

Homeopathy is an exact medical science; proponents claim it is a "natural" system of medicine that, again, treats the whole person, uses no toxic drugs, and in many cases, cures "hopeless" chronic conditions. For more on the subject, you may want to read *Homeopathic Medicine at Home* by Maesimund B. Panos and Jane Heimlich, and *Everybody's Guide to Homeopathy* by Dana Ullman, M.P.H. and Stephen Cummings, F.N.P.

Naturopathic. The word literally means "nature cure." Naturopathy embraces such therapeutic methods as fasting, hydrotherapy, massage, vitamin and mineral therapy, nutrition supplements and special diets, herbal treatments, spine and joint manipulation, acupuncture, hypnotherapy, biofeedback, and colonic irrigation.

The approach is holistic; practitioners emphasize the importance of diet, fresh air, exercise, and peace of mind, and integrate prevention, education, and psychological counseling in their treatment.

Naturopaths are not M.D.s but receive a Naturopathic Doctor degree after four years of study at a specialized college. "We're more into preventive rather than crisis medicine," says Irvin H. Miller, N.D., president of the National Association of Naturopathic Physicians in Tacoma, Washington. "We treat patients with chronic conditions such as arthritis, hypoglycemia, allergies, obesity, fatigue, and more. But we recognize there are conditions that we can't handle, and we refer those patients to M.D.s and D.O.s."

The legal status of the naturopath varies widely; some states prohibit the practice, other states grant licenses and enforce regulation.

Orthomolecular. The word "orthomolecular" was coined by Nobel Prize-winner Dr. Linus Pauling and literally means "correcting molecules"—in this case, restoring the various systems to optional functioning by prescribing substances normally present in the body and healing with nutrition and megavitamins.

Orthomolecular doctors can be dentists, chiropractors, D.O.s, or M.D.s, and often see patients who have "gone the rounds" with traditional doctors. Some people have been helped; others have spent huge amounts of money on vitamin and mineral supplements and have shown no improvement, or in some cases, neglected the orthodox medical care they needed and become worse.

Over the centuries, however, many people have been successfully treated by alternative medical approaches. If you decide to try one, choose a reputable practitioner, keep an open mind, and use all prescribed products in moderation.

Recreational Drugs

There are a few alternatives to alcohol, and many substitutes for caffeine beverages and nicotine, but unfortunately, no effective nontoxic substitutes for marijuana and cocaine. Tetrahydrocannabinol (THC), the active ingredient in marijuana, is frequently sold on the streets as a "pure" intoxicant, but since the process of extracting THC from the cannabis plant is very costly, buyers are not getting THC at all. Usually, they will be sold the extremely dangerous phencyclidine—PCP or "angel dust."

Some people claim that tea made from Gotu Kola, a Fijian herb, sedates the part of the brain affected in cocaine withdrawal and serves as a weaning device. Others tout Sudafed, a popular over-the-counter decongestant, as an alternative stimulant to be taken during the first three days of cocaine withdrawal. Neither has been proven to be effective, and both could be harmful.

Alcohol. Healthy substitutes for alcoholic beverages, labeled "de-alcoholized" drinks, have recently become available in American gourmet food shops. Unlike the carbonated grape juice and half-brewed beers of yesteryear, the new de-alcoholized drinks are fully brewed or fermented before all but 0.5 percent of their alcohol is removed. The taste and body of these beverages is very similar—sometimes indistinguishable—from the real thing. An added bonus is that they contain two-thirds fewer calories.

In late 1982, Alan Luks, executive director of New York's National Council on Alcoholism, tested two de-alcoholized white wines and reported that one was "too sweet and fruity," but "I enjoyed Giovane, a sparkling Italian wine. Both had a distinct advantage over regular wine; without alcohol's distortion of the senses, I found I enjoyed my food far more."

In a later test, Luks tried two cans of de-alcoholized beer, and concluded, "They offered the same refreshing taste of regular beer. The foamy beverages also left my stomach with the fullness of beer. I felt unusual—then I realized that I was noticing the lack of any effect on my thinking. My mind felt so strangely clear."

Two main problems exist. One is that substitutes for distilled spirits such as gin, vodka, and bourbon are not yet perfected, and two is that recovered alcoholics, accustomed to avoiding desserts, salad dressings, and other foods with even a trace of alcohol, should also avoid the 0.5 percent in de-alcoholized wines and beers. The Alcoholics Anonymous publication, *Grapevine*, reported in November 1983 that a Chicago chemist was trying to perfect a low-calorie bourbon substitute. The editorial hinted that the man should save his time and money, and explained: "The only answer for an alcoholic is No-Booze Booze."

Nevertheless, these new de-alcoholized wines and beers provide the first real alternative to the overuse of alcohol, and if properly marketed and accepted by the public, may do for wine and beer what decaffeination has done for coffee.

Caffeine. Thirty million Americans consume more than two cups of decaffeinated coffee every day, and the numbers are growing. New techniques for nonchemical extraction of caffeine—notably, the "water process"—make the brew safer, tastier, and more popular than ever.

For those who do not even want the 2 to 4 milligrams of caffeine in a cup of decaffeinated coffee, there are nutritious alternatives. Check the ingredient list to make sure the substance you are avoiding is truly absent and to ascertain that you are not drinking other toxic substances.

Pero, Pionier, Postum, Cafix, and Duram are coffee substitutes made from roasted barley, rye, chicory, and shredded beet roots. They contain no coffee or caffeine, and their taste may take some getting used to, but many people are fans. Some researchers warn that any beverage made from *roasted* grains may be carcinogenic, but unless a person drinks ten or more cups a day, the dangers appear to be negligible.

Carob drinks such as Cara Coa and Caroba are instant drinks with a taste similar to chocolate. They contain no caffeine but are usually sweetened with fructose or some form of sugar.

Herbal teas are made from plants other than the *Camellia sinensis*, which produces both green and black tea. Beverages listing black tea, green tea, or matte (also spelled mate) contain caffeine and ought to be shunned.

Herbal teas come in many brands and flavors and should be rotated for variety and to avoid building up sensitivities or suffering harmful effects. Dr. Bruce Ames, chairman of the biochemistry department at the University of California in Berkeley, warns us to be wary of such herbs as buckthorn, burdock root, chamomile, ginseng, jimson weed, juniper berries, mistletoe, nutmeg, pokeweed, sassafras, senna, and shavegrass. People have reported severe diarrhea from buckthorn, hallucinations from burdock and jimson weed, and stomach cramps from juniper berries, mistletoe, and nutmeg. Mint and chamomile can stimulate uterine contractions and threaten miscarriage; ginseng has been reported to race the heart and contains small amounts of estrogens, which can cause swollen breasts.

Currently recommended as a "safe" coffee substitute, Chi Power is a blend of three Chinese herbs: astragalus, claimed to be an immune-system enhancer; polygonum multiflorum, a rejuvenator; and ephedra, which contains the decongestant and excitant ephedrine. Chi Power is a potent stimulant that should be used with caution, and never with other stimulants or by anyone with heart problems or hypertension. Those who market it claim it supplies a non-caffeine lift for late-night studying, but a writer in *Esquire*, March 1984, tried it and reported, "It was such a crude, undifferentiated surge of energy that it was impossible to focus on anyone or anything. My pulse was pounding, my brain was buzzing, my mouth was dry. When the 'high' was over . . . the hangover began."

Some safe teas are red zinger, lemon balm, anise, rose hip, raspberry, and lemon grass. All the new boxes of Celestial Seasonings tea, including the pleasant Mandarin Orange Spice, clearly proclaim NO CAFFEINE on the package. (Morning Thunder, the sole exception, contains 45 milligrams of caffeine per cup.) If in doubt about the effects of any tea, ask your doctor, write the company that markets it, or consult a reputable herbalist. Another possibility is "white tea"—plain boiled water with a slice of lime or lemon.

Many people who take pain relievers with caffeine, such as certain brands of aspirin, might do better taking the same pain reliever without caffeine. Read the labels on all painkilling pills, cold remedies, and other nonprescription drugs. Although a 1984 report in the *Journal of the American Medical Association* indicated that caffeine boosts the action of pain relievers, its central nervous system effects can make you feel jumpy just when your body most needs the healing powers of sleep. Brands without caffeine are almost as effective; no medication at all is best.

Nicotine. Food, especially sweets, has long been a handy alternative to smoking, but few people want the cavity potential or the empty calories. Low-tar cigarettes have been proven to be no less carcinogenic than their high-tar cousins, and pipes and cigars, if inhaled, are as harmful as cigarettes. None of the above is a satisfactory alternative to smoking, but there are some other possibilities.

Dry snuff is a powdered form of tobacco that is sniffed, not inhaled. Its advantages are that it provides nicotine without contaminating the air, contains no tar and does not harm the lungs, gives the smoker something to do with his hands, and is free of carbon monoxide and other gases in smoke that irritate the heart. Snuff comes in a variety of flavors, including menthol. Its best use is to help a smoker cut down before quitting.

Moist snuff is placed in a small pinch in a corner of the user's mouth. It stays there, does not interfere with talking, and is absorbed through the mucous membranes.

Chewing tobacco is a bit coarser than snuff but can serve the same purpose—a weaning device between smoking and stopping.

The major disadvantage with all of these alternatives is that they maintain the blood nicotine level, prolong addiction, and postpone withdrawal. In September 1983, Dr. William Regelson of the Medical College of Virginia revealed new research indicating that snuff and chewing tobacco contain potent cancer-causing agents. While still preferable to smoking, these are temporary aids at best.

Other weaning devices include over-the-counter preparations such as Bantron and Nicoban that contain a plant derivative, lobeline sulfate, which has stimulant and euphoric properties similar to nicotine but is less toxic. The newest product is in the form of chewing gum laced with nicotine. Dr. M. A. H. Russell of Maudsley Hospital in London made a study of 116 smokers. Half were given nicotine gum, and half received gum without it. All were asked to chew whenever they felt the urge to smoke.

After a year, 49 percent of those who chewed nicotine gum had quit and stayed off cigarettes, as compared with 19 percent who used the placebo. A newer study reported in the 1983 *British Medical Journal* tested 1,938 smokers and found equally convincing results.

Dr. Russell recommends switching immediately from cigarettes to gum, chewing it for four months, then slowly tapering off. A supply of gum should be kept on hand for at least a year, in case the person is ever tempted to reach for a cigarette.

Nicorette, as the gum is called, is available by prescription only and should not be used by pregnant women.

Strictly for oral gratification, why not try:

Hard candy. One woman reported, "I wanted to kill the pangs by 'chain-sucking' tart lemon drops, but all I did was ulcerate the inside of my mouth and get six cavities." Use candy on a temporary basis only.

Ice cubes. If the cold does not bother you, these are good to chew and crunch, and may have a slightly anesthetizing effect. Don't bite down too hard, however, as you could chip a tooth.

Low-cal munch foods. Keep apples, nuts, and carrot and celery sticks handy.

Manual props. Try using a fake cigarette, fiddling with a toothpick, or massaging your gums with a rubber prodder.

Filters. Products such as Water-Pik and Venturi work to gradually decrease the tar and nicotine content of your cigarettes, so that after using filters for four to eight weeks, as much as 95 percent will have been removed.

Mouthwashes. Sold as the smoker's answer to Antabuse, these special gargles contain chemicals that mix with smoke and leave an unpleasant taste. A few people have found them useful.

Spices. It may help cravings to suck on small pieces of whole clove, cinnamon, or ginger, although some people say they "burn" the tongue.

Sugarless gum. Preferable to candy, gum chewing may help, but is physically unattractive at a time when you need to attract understanding and support.

A smooth pebble. Suck on it, roll it over your tongue, and have no worries about calories or cavities—but don't swallow it.

A baby pacifier. You won't want to nibble this in the company car pool, but it might comfort you in the privacy of your bedroom.

Having started your General Detox Program, you have officially alerted your body that help is coming, in the form of diet, exercise, and relaxation routines. With any change of habit, it is wise to anticipate withdrawal symptoms. In this next chapter, specific programs for detoxifying yourself of individual substances will carry you through the physiological and emotional symptoms you may experience.

5

Addiction and Withdrawal: Escaping Your Chemical Captors

The greatest hurdle most people face when detoxing from habit-forming substances, such as nicotine, alcohol, or cocaine, is going through withdrawal—with its variety of physiological and emotional symptoms—unscathed. Our guidelines will prepare you psychologically for those difficult moments when your body's chemistry is aching for the burst of energy from a cup of coffee or the calming effects of a tranquilizer. We explain how to tell if you are addicted and how to identify withdrawal symptoms, and describe specific detoxification procedures for each of the following: alcohol, caffeine, chemicals, cocaine, foods, sugar, marijuana, medication, and nicotine. Information on how to find professional help is also included.

HOW TO KNOW IF YOU ARE ADDICTED

You may not think you are addicted—few people do. If, however, under normal circumstances, you cannot control when you start or stop an activity, consider yourself addicted.

Let us look at three people who claimed they could easily break their habits.

Case 1. Gary B. had frosted corn flakes for breakfast every morning, drank sugar-drenched Sanka on his coffee break, lunched on peanut butter, jelly, and whole wheat bread, snacked on candied fruit bars, washed down dinner with sweetened herbal tea, and topped the meal with carrot cake. Despite taking a variety of vitamin pills and shopping at health food stores, Gary felt anything but healthy, with constant anxiety and irritability, and frequent headaches. By the time he was forced to admit to himself that he was a sugar junkie, Gary was into his second year of addiction.

Working as a customs officer often involved physical exertion; Gary feared he would not get through the day without the energy he got from sugar. He decided to start kicking his habit by cutting down rather than cutting out. First to go were the between-meal snacks; his Sanka tasted fine with saccharin, and while fresh fruit was less satisfying, it was at least a filling replacement for the candy bars.

Then he decided to eliminate all sugar from his diet at once, replacing it with nuts, seeds, and fruits, and aiding the detox process with large quantities of water. He felt some discomfort for three days and suffered occasional dizzy spells, blurred vision, and periods of depression. He had little energy for work, but his boss would hardly understand if he took time off to "withdraw from candy bars."

On the fourth day, his depression cleared, his mind seemed strangely sharp, his palate had recovered its blunted sensitivity, and the natural flavors of food almost overwhelmed his taste buds. Gradually his physical energy returned, and with it, a sense of self-mastery and control. Now he is an ex–sugar junkie and proud of it. Life is so much sweeter.

Case 2. Patsy F., a social worker, smoked three packs of cigarettes a day. Though her husband and children nagged her to quit, she knew that she was not addicted and could stop at any time with no problems.

Patsy always bought cigarettes at the office vending machine. One day the price was hiked, and she asked a coworker to lend her the extra quarter. The woman did so, growling, "You'll pay anything to poison your lungs!"

"Her remark hit me," said Patsy. "If cigarettes had cost $25 a pack, I would have paid it. When I realized how far gone I was, I turned and walked away from the machine. It was murder that first week; I had a headache that wouldn't quit, stomach cramps, and muscle aches. I wasn't sure I'd come out of the depression, but I did—and I haven't had a puff since then."

Case 3. Mara D., a young fashion model, used a variety of cosmetics, but only one kind of hair spray: Red Rose. It kept her locks in place, scented the air, and gave her a psychological lift. Unknown to her, the toxins stimulated her central nervous system and supplied a physiological lift as well. Mara bought Red Rose from the druggist in case lots, always carried a can in her purse, and used it four and five times a day.

After many months, a constant runny nose and physical exhaustion led her to Dr. Saifer, who sniffed her aura of roses and became suspicious. Mara's strong denial of a possible link between her symptoms and her hair spray led her to see an old-fashioned physician who prescribed antihistamines, tranquilizers, and a vacation. When nothing helped, she decided to try a day without Red Rose—and felt so miserable, she finally began to realize her chemical dependency.

Returning to Dr. Saifer, she took the physician's advice to discard the Red Rose and restyle her hair into a casual bob that stayed in place by itself. After a week of irritability, frustration, and aching muscles, her nose stopped running, her fatigue disappeared, and her good spirits and youthful energy came bouncing back.

The fact that all three people were able to effect their own withdrawals without drugs or hospitalization does not mean that they were not hooked. Dr. Stanton Peele, psychologist and author of several books on addiction, interviewed subjects for more than ten years and found "intriguing evidence . . . that people who break bad habits do best if they do it themselves. Heroin addicts often quit use on their own. Alcoholics frequently don't need to dry out in a hospital but just go on the wagon with no particular anguish. Practically every cigarette smoker stops at some point—for anywhere from a few days to years."

Look around you. Are there chemical substances you eat, drink, wear, inhale, or spread on your skin that would be very difficult to part with? Can you give up any or all of them without anguish? Is there one you would cling to over all the others? If you have tried doing without the substance before and found it impossible, then you are addicted.

A more drastic means of finding out if you are addicted is to cease doing whatever you are doing. (Check first with your doctor if you have been using the substance for more than a year.) When you eliminate the habit, watch for possible symptoms, which may occur hours or up to a week after your last contact with the offending substance, that may indicate the beginning pangs of withdrawal.

Withdrawal symptoms can range in intensity and severity from mild anxiety and irritability to blackouts and seizures, although these latter are rare. Successfully coping with withdrawal and breaking your habit depends on:

1. Age and general health. Younger people usually have "newer" habits and more resilience. A healthy body offers more physical resources to draw on.

2. Mental state and psychological stress load. A positive attitude and freedom from tension are prerequisites to any successful withdrawal.

3. The length of time addicted. The shorter the addiction, the easier it is to break.

4. The nature of the substance, be it bubble bath or

barbiturates. Most people find caffeine less addictive than cocaine, and so on.

5. The dosage or concentration of the substance; if medication, the spacing of dosages and route of administration. Pills and tablets provide a problem because they are so handy and so easy to swallow.

6. The availability and extent of medical care, support groups, and family assistance, if needed. Human aid is never more than a phone call away.

7. Whether the addiction is private, or part of a subculture involving peer group pressure to continue use. So-called "friends" often contribute to the problem.

8. Personal habits, including use of other drugs or toxic substances. Reduce the amount of synthetic chemicals in your life to hasten your recovery.

HOW TO IDENTIFY WITHDRAWAL SYMPTOMS

People often confuse withdrawal symptoms with emotional stresses or other body malfunctions. A headache, for instance, will almost always be the first sign of quitting caffeine use, yet it can have hundreds of other sources.

An effective method of identifying a withdrawal pattern is to consider the factors listed above, such as dosage and length of time used, and group them together with all mental and physical symptoms. Then, check your responses against the table; for instance, if you have stopped smoking and are feeling depressed, nauseated, and have a headache, you will see that nicotine addiction (number 8 on the table) matches those symptoms, so you can safely assume you are undergoing nicotine withdrawal.

You need not experience all the reactions listed for a particular substance to confirm your diagnosis; one symptom alone is enough to indicate you are undergoing withdrawal. In general, milder reactions indicate less severe addiction.

Withdrawal Symptoms

Numbers 1 to 9 represent the addictive substances and appear after the withdrawal symptoms they usually produce. An asterisk indicates the likely need for medical supervision; two asterisks mean a doctor's care is essential. If you suffer any acute symptom longer than twelve hours, or if it causes more discomfort than you can handle, call your doctor immediately.

1. Alcohol
2. Caffeine
3. Chemicals (in cosmetics and toiletries, inks, paints, glues, and other products)
4. Cocaine
5. Foods
6. Marijuana
7. Medication; all types
8. Nicotine
9. Sugar

Likely

Unlikely

Head

Muscle tension headache; 2, 3, 5, 7, 8, 9
Migraine; 2

Headache associated with stiff neck, fever, numbness anywhere on body

Eyes

Blurred vision*; 5, 7, 9
Dilated pupils*; 1, 7
Double vision*; 7

Flashing lights and visual disturbance preceding migraine

Likely	*Unlikely*
Rapid side-to-side movement of eyeballs*; 1, 7	Itching and redness with pain
Uncontrolled winking*; 7	
Watery eyes*; 7	Visual problems in one eye only

Nose

Runny nose; 2, 4, 7	Thick yellowish discharge
Smelling unpleasant odors; 7	
Nasal tissue irritation; 4	

Ear

Ringing in ears; 2	Pain and itching with deafness
Ache in ears; 7	
	Pain in one ear only
	Deafness, dizziness, with tendency to fall in one direction

Mouth and Throat

Bad taste; 7	Canker sores
Dry mouth; 1	Gum infection
Sore gums or tongue; 8	Pain in single tooth
Yawning; 7	Difficulty swallowing

Chest

Irregular heartbeat*; 7	Irregular heartbeat associated with history of cardiac problems
Rapid heartbeat*; 2, 3, 9	

Likely

Difficulty breathing*; 7

Stomach and Intestines

Constipation*; 7, 8
Diarrhea*; 2, 5, 7, 8
Nausea and vomiting*; 1, 2, 5, 7, 8
Stomachache, cramps*; 2, 3, 5, 7, 8

Genitourinary Tract

Incontinence*; 7
Urinary frequency*; 2

Skin

Flushing; 2
Gooseflesh; 7
Hives all over body; 3, 5
Itching; 3, 5
Psoriasis; 3, 5
Rash; 7

Brain Effects

Anxiety; 1–9
Apathy; 2, 4, 7

Unlikely

Cough producing blood or sputum

Pain that starts below breastbone and radiates to back or right side of ribs
Pain that appears mid-morning, is relieved by food, then recurs two to three hours after a meal

Genital rash with itching and discharge
Changes of urinary habits with fever, weight loss, and pain

Rash weeping yellow discharge or pus

Severe burn on exposure to small amount of sun

Changes of pigmentation

Blue or yellowish color

Emotional problems or behavior related to specific event

Likely

Craving of addictive sub-
stance*; 1–9
Delirium**; 1–9
Depression*; 1–9
Dizziness*: 1, 2, 3, 5, 7, 9
Drowsiness; 2, 4, 8
Fatigue; 1, 4, 5, 7
Hallucinations**; 1, 4, 5, 7
Hyperactivity; 1, 3, 5, 6, 7,
9
Inability to concentrate; 1,
2, 3, 4, 8, 9
Insomnia; 1, 2, 4, 6, 7, 8
Irritability; 2, 4, 5, 7, 8, 9
Loss of appetite; 1, 6, 7
Nightmares; 4, 7
Panic**; 7
Rage reaction*; 7, 9
Seizures**; 7
Sensation of insects crawl-
ing on skin**; 7
Temporary insanity**; 1, 3,
4, 5, 7
Tension; 1, 2, 3
Unsteady gait; 1, 2

Unlikely

Craving linked to
pregnancy

Personality change with fever, headache, vomiting,
and stiff neck

Mental changes with
numbness on one side of
body

Dizziness associated with
painful earaches

Inattention, lack of motivation, with headache, loss
of limb control, flashing
lights.

General

Weight gain; 4, 8
Weight loss; 7
Chills; 1, 2, 5, 7
Dehydration; 1
Fever and sweating**; 1, 2,
5, 7

Chills with bloating, numbness in extremities

Likely	Unlikely
Hiccups; 7	
Lower back pain; 7	Lower back pain radiating
Muscle aches; 3, 4, 7, 8	down one leg only
Muscle twitching; 4, 7	
Muscle weakness; 1, 7	Muscle weakness with dra-
Tremors*; 1, 2, 5, 7, 9	matic visual changes
Weakness; 1, 2, 3, 4, 7, 9	

THE INDIVIDUAL DETOX PROCEDURES

Now that you have determined the range of your withdrawal symptoms, there are precise steps you can follow to keep them at a minimum. Every addictive substance presents a unique set of challenges with its own rules and remedies. The desired result of these specific detox programs is to achieve complete freedom from addiction—psychologically as well as physiologically.

Alcohol

Both heavy and moderate drinkers are choosing to live the rest of their lives without alcohol. Their decision is based on several factors: the potential for abuse of this drug; its empty calories leading frequently to malnutrition; toxic effects on the brain, heart, muscles, liver, and nervous system; impairment of mental faculties; and even the waste of time and energy the cocktail circuit demands.

The short-term problems of giving up alcohol usually center on fear of social and job-related rejection, maintaining one's motivation, withstanding outside pressures to drink, and worries about withdrawal. These stumbling blocks can all be removed with care and education.

Long-term benefits appear in many forms: improved nutrition and vitality; better health and chances for longevity; renewed self-confidence and self-respect; regained mental acuity; time for friends, family, and interests; happier personal relationships; weight loss or gain—whichever is desirable; and no more bar or liquor bills.

Although experts disagree on whether or not reformed alcoholics can be taught to drink socially, most feel that the safest path is abstinence. For our purposes, alcohol is a potent drug with many toxic properties and of little or no therapeutic value.

Alcohol Detox Procedure

1. Admit that you have a problem—either with "social drinking" or with addiction.

2. Remove all the liquor in your home, your car, your desk at the office, and wherever else you store it.

3. Tell your family and friends. Announce your new non-drinking status to everyone. It will strengthen your resolve to continue and make you reluctant to appear a failure in the eyes of others.

4. Consult your family doctor and get a physical examination if you have not had one in the last three months. If he suggests sending you to a clinic or hospital, and you feel it is unnecessary, discuss alternatives. If your doctor recommends a mild tranquilizer to ease the withdrawal symptoms listed in step 6, take it only if your discomfort is severe; drugs add to the toxicity buildup. Be sure to mention any medication you are already taking, as the dosage may change.

5. Consider the possibility that your dependency may be wholly or partly caused by an allergic reaction to a food or a chemical. (This is explained in detail later in the chapter.) The person who wants a particular brand of beer and no other brand may actually be craving the corn or rice in the brew rather than the alcohol. If you have an allergic condition, or suspect you have and wish to get tested, see a

clinical ecologist—an allergist trained in food and chemical sensitivities. Clinical ecologists only treat allergies and help the mystified person identify the culprits; they do not treat alcoholism, drug dependency, or any severe addiction problem.

To get the names of clinical ecologists in your area, write to the Society for Clinical Ecology, which is listed in Appendix L.

6. If your doctor does not recommend medical treatment, or you think you can handle the situation alone and without outside help, skip to step 8. If your doctor recommends medical supervision, or you feel you need some type of assistance, you have several choices of treatment to ease you through the four stages of alcohol withdrawal. You may experience some, none, or all of these symptoms:

Tremors, sleeplessness, nausea, irritability, anxiety, shame, guilt, a sense of inferiority, depression, and a craving for alcohol within six to eight hours after the last drink.

Hallucinations of sight, smell, sound, and touch within twenty-four hours.

Although not common, a few people will have two to six brief grand mal (major epileptic) seizures in an eight-hour period, eight to forty-eight hours later. If protected from injury and aspiration—the sucking of solids or fluids into the windpipe— the patient rarely suffers harm. The person having the seizure should *not* be given anything to eat or drink, and nothing should be placed between his teeth. Clothing should be loosened and medical care summoned immediately. Again, this reaction is possible, but extreme and unusual.

Delirium tremens with its trembling, feelings of persecution, hallucinations, and ultimate exhaustion may occur any time from seventy-two hours to fourteen days after the last drink.

If you have been drinking heavily—that is, eight or more ounces of alcohol a day—and have been doing so for six months or more, plan on at least three days of detoxification under medical supervision with residential care. If you have been drinking lesser amounts, or have successfully abstained before without live-in care, you may do just as well in a nonresidential center.

Nonresidential clinics offer two to eight hours a day of some or all of the therapies listed. Residential clinics and hospitals offer three- to thirty-day live-in programs and provide seven to twelve hours a day of:

Alcoholics Anonymous orientation and meetings, if desired

Alternatives to drinking

Attitude and assertive-behavior training

Educational lectures and films

Family counseling

Goal-setting discussions

Individual and group therapy

Individual and group counseling

Medication and medical care

Meditation and relaxation instruction

Planning of aftercare

Religious inspiration, if desired.

Almost all residential clinics provide a continuous monitoring of vital signs, emergency medical care, and medication in the form of tranquilizers or sedatives—usually Valium, Serax, Librium, or Phenobarbital—to ease withdrawal. Hospitalization has several other advantages: It removes you from your environment and the familiar conditions that trigger your drinking; it emphasizes the seriousness and disease nature of the condition; and it makes your decision to quit an important commitment, rather than a haphazard fling at sobriety.

7. Realize that what works for one person will not necessarily work for you. Each treatment regimen must be individually tailored for each patient. Below are several alternative therapies available at some residential and nonresidential centers.

Acupuncture. This treatment consists of inserting hair-like needles in the ear rims, lobes, hands, or legs, at points that correspond to the lungs, liver, and kidneys—major pathways for the excretion of toxins. Dr. Michael O. Smith of the alcohol and acupuncture program at Lincoln Hospital in New York explains that acupuncture stimulates the body's own detox system to work faster and more effectively and also exerts "an overall calming effect."

Most patients start with four sessions the first week, two the second week, then one a week for one to three months. Sessions last twenty to sixty minutes, are either painless or cause minor discomfort, cost $35 or more, and are not usually covered by medical insurance.

Antabuse. This very powerful drug causes violent nausea and vomiting when taken with alcohol. The patient takes an Antabuse tablet, then refrains from drinking to avoid the pain and discomfort. Antabuse may also induce such side effects as deep sweating, a large drop in blood pressure, chest pains, irregular heartbeat, and psychotic behavior, and for these reasons is used mainly when other programs fail.

Current research at the University of Minnesota centers around cyanimide, a compound which creates the same aversion to alcohol as Antabuse, but does not cause the severe side effects. At present, cyanimide is awaiting FDA approval.

Glutamine. Twenty-five years ago, nutritionist and chemistry professor Dr. Roger Williams began recommending glutamine—an amino acid found in liver, cabbage, and high-protein foods such as meat, fish, and dairy products—to help restore the nutritional imbalance that causes alcoholic craving.

Glutamine (1-glutamine or glutamic acid) is available in powder form in health-food stores. Dr. Williams advises taking half a gram—about one-eighth of a level teaspoonful—before and after meals and before bedtime. "It's a nutrient, not a drug," he says, "and for many, it curbs the unhealthy drive to drink. It should be used as long as it appears to be effective."

Lithium. Widely prescribed for manic-depressive disorders, lithium may be the first biologically effective medication for alcoholic craving. Dr. Jan A. Fawcett of Rush-Presbyterian-St. Luke's Medical Center in Chicago, states that, "Lithium does not affect depression, but acts by *suppressing* the urge to drink." A 1983 Rush study involved eighty-four patients who had abused alcohol for an average of seventeen years. Seventy-five percent of the drinkers who took daily lithium pills during their initial abstinence, as well as afterward, were still abstinent eighteen months later, 100 percent of those who stopped taking the drug had resumed drinking by the end of six months.

Nutrition. A growing number of researchers feel that more attention must be paid to the biochemical factors that create alcoholism, rather than simply treating it as an emotional disorder. Joan Mathews-Larson, nutritionist-director of the Health Recovery Center in Minneapolis, says, "People heal very nicely here without confessing their sins in group therapy sessions and without being made to feel ashamed of their disease." Treatment at the center includes avoidance of coffee, tobacco, sugar, and white flour; testing to unmask food allergies that could cause alcoholic craving; and daily nutritional supplements.

Harry K. Panjwani, M.D., former board member of the National Council on Alcoholism, also reports good results by starting new patients on a regimen of glutamine, vitamin B3 (niacin), and vitamin C—at least 500 milligrams of each a day. Britain's Dr. David Horrobin recommends taking the above along with at least four 0.5 gram capsules of evening primrose oil to help block withdrawal symptoms and prevent craving.

8. Once you set down your last glass, start changing your habits. Avoid bars, parties, drinking chums. Learn to decline with pride and grace. Never feel that you owe anyone an explanation. "No thanks, I'm not drinking," is all you need to say. Better yet, order soda or Perrier water with a slice of lime and say nothing. You may be surprised to find that the same society that once derided the nondrinker as a bad sport, or not "one of the gang," now respects and supports your concern for your body.

9. Change your routine to avoid temptation. If you had a cocktail every evening at six, get out of the house, take a walk, visit nondrinking friends, go for a drive, or go to a movie.

10. Take steps to maintain your detox regimen. Alcohol withdrawal actually marks the beginning—not the end—of rehabilitation. Both residential and nonresidential centers usually include a year of aftercare for you and your family.

11. Lean on your family or friends. By now they should be aware of how hard you are trying, and possibly have been educated in the best ways to offer support. Two weeks to two months after alcohol detox can be the most stressful time. Your sense of reality is no longer distorted, and you have to face all the problems you have been blocking out. The tendency to want to retreat back into alcohol can be strong.

12. Do not despair if you slip. An occasional setback does not mean that you are a failure or that you have taken on an impossible task. Use the experience to strengthen your resolve not to do it again.

13. Spoil yourself. Indulge in other ways. Use the money you save not buying alcohol to enjoy a cruise to China, take

up ballooning, learn basket weaving, or buy a computer—with private lessons! Reward yourself in every possible way so that your efforts add up to a more fulfilling life.

Caffeine

Marilyn B., an advertising executive, was drinking twelve to fourteen cups of coffee a day when she went to her doctor for a physical, complaining of "raw nerves" and frequent anxiety attacks. Neglecting to take a diet history, the doctor prescribed the tranquilizer Librium and psychotherapy.

Fortunately, Marilyn's boyfriend was a health enthusiast with a passionate hatred for all drugs. He suspected the problem was caffeine addiction and begged her to first try tapering off. She did so, experiencing some flushing and sweating and a mild headache the first week, but no symptoms at all the second week, when she switched to a decaffeinated brand.

Two weeks from the day Marilyn started cutting her caffeine intake, her anxiety disappeared, her nerves "healed," and she reported "a most amazing rejuvenation." Coffee still sits on the kitchen shelf for guests, but taped to the can is a poignant reminder: the unfilled prescription for Librium.

In moderate doses, caffeine probably does no more harm than the sugared coffee, soft drinks, regular tea, chocolate, pain relievers, "water pills," and diet aids that contain it. The moment your intake begins to bring on such symptoms as craving, jumpiness, insomnia, headaches, anxiety, and irritability, however, you have reached the danger level. You are taking too much.

Caffeine Detox Procedure

1. Figure out your daily caffeine consumption. The average adult gets a lift from 150 to 250 milligrams. These approximate figures show how much caffeine there is in a standard five-ounce cup of each type of coffee:

Drip, 150 milligrams

Percolated, 110 milligrams

Instant, 60 milligrams

Decaffeinated brewed, 4.5 milligrams

Decaffeinated instant, 2 milligrams.

Twelve-ounce soft drinks such as Pepsi Cola, Diet Pepsi, Coca-Cola, Tab, and Dr. Pepper contain 35 to 60 milligrams caffeine; a cup of regular tea supplies 45 milligrams; hot cocoa has 13 milligrams; one ounce of chocolate has only 6 milligrams.

A single Vivarin, Dexatrim, or Codexin tablet contains 200 milligrams; NoDoz and Aquaban have 100 milligrams; Excedrin has 65 milligrams; and Anacin, Midol, Vanquish and Cope tablets each have about 30 milligrams.

2. Keep a journal of how much coffee you drink, as well as all noncoffee sources of caffeine; physical dependence can occur on five or more cups a day. Every evening for a week, add up and write down your total intake. More than 500 milligrams a day is considered heavy usage.

3. Taper off gradually. Mix half-regular and half-decaffeinated coffee. Cut pills into halves or quarters; reduce your soft drink intake. Eliminate 100 milligrams a day for 3 days, another 100 milligrams for the next 3 days, and so on, down to zero. Simply changing from drip to instant saves 100 milligrams a cup. Switch to decaffeinated and your intake drops 50 milligrams more.

4. Since you are tapering rather than stopping cold turkey, it is unlikely that you will get severe withdrawal symptoms. Quitting should provide almost immediate relief from jangled nerves, yet the possibility exists that you may experience one or more of the following: headache, tremors, irritability, tiredness, lethargy, anxiety, or depression. Any symptom can last from one day to two weeks. Avoid exercise and heavy work which will aggravate a headache; sleep

when you feel drowsy; and if concentration is difficult, save your detailed work for two weeks hence.

5. Some tricks may help: change your mug for a smaller cup; measure a level, not heaping teaspoon of coffee; pour container only two-thirds full; sip brew slowly or with a spoon; dilute it with hot water or low-fat milk.

6. Dr. Charles Ehret of the Argonne National Laboratory near Chicago urges users who get down to one cup a day, to drink that cup at about teatime—4:00 P.M. Studies of body cycles show that coffee will do the most to perk you up and have the least upsetting effect on your central nervous system at that hour.

7. Check the alternatives listed in Chapter 4 and proceed with caution. Caffeine-free soft drinks may be just that, but the plethora of chemicals in these concoctions can be as bad or worse for you. Decaffeinated coffee still comes from beans sprayed with pesticides, and the caffeine may have been extracted by chemical solvents of questionable safety. The ingredients list of nonprescription pain relievers demands careful scrutiny, as it may contain other dangerous or addictive substances.

An investment counselor, Karl P., made up his mind to quit his twenty-cup-a-day habit cold turkey. He chose Friday after work as the best time to stop; he did not need the lift on Saturday morning, and if he felt poorly, at least he had the weekend to convalesce.

Karl awoke Saturday with a severe headache, took two aspirin, and felt worse. An hour later, he took two Anacin tablets and his head miraculously cleared. He found that two Anacin every few hours kept him feeling fine, but by Sunday, he realized the reason: The caffeine in the Anacin simply continued his addiction and delayed the withdrawal.

He then had to taper off the tablets, spacing them further and further apart. He experienced only a mild headache.

Two weeks later, he was taking neither Anacin nor coffee and feeling calmer than he had in many months.

8. Abstaining from caffeine has become socially acceptable, yet some people may still call you a health nut or an oddball. Your calm nerves, lack of anxiety, ability to sleep more soundly, and improved well-being should be adequate compensation for any teasing.

9. To maintain your detox condition, never indulge yourself with "just one little cup." Coffee is so widely served and accepted, it takes exceptional willpower not to fall back into the same old addictive pattern.

Chemicals

Dr. Theron Randolph of Chicago was one of the first medical practitioners to observe that some of his patients with chemical sensitivities tended to seek out the very substances that were harming them. They had even come to depend on continued exposure to keep their bodies in an altered state of homeostasis or body balance. Once the substances were removed, the metabolic functions readjusted in such ways as to cause both mild and severe temporary withdrawal symptoms.

The printer who carries fresh samples of his work wherever he goes, the beautician who feels terrible on weekends, and the artist who is compelled to go to his studio seven days a week may all be as hooked on chemicals as a compulsive glue sniffer. The printer is addicted to ink fumes, the hairstylist has to inhale scents and sprays, and the artist must have his quota of paint solvents in order to maintain a feeling of well-being; otherwise, all three will experience withdrawal.

George J. worked for a commercial exterminator, and although he wore a cotton mask to spray insecticide, he had never been too careful about his hands, hair, or clothes. Over the years he noticed that he felt more and more sluggish in the early morning and on days when he only did paperwork.

The moment he grabbed his overalls, spray gun, and other supplies his energy level jumped. The pattern became so predictable, and he felt so poorly between exterminations, he suspected that something abnormal was happening in his body.

A friend sent George to see Dr. Saifer, who tested him and found a marked sensitivity to kerosene, a common petroleum distillate used in insecticide sprays. When George learned that he was addicted, he decided to go to an outpatient detox center. He suffered dizziness, muscle cramps, and a headache for the first two days. On day three, he reported a gassy taste in his mouth, a strong kerosene odor about his body, and an intense flashback, in which he reexperienced a long-forgotten illness. Afterward, he slept for twelve hours.

The headaches cleared on the fifth day, and by week's end, all his symptoms had disappeared. He felt shaky but encouraged, and gradually regained his strength at home. Luckily, he was able to change chores at work, so that he no longer had to spray insecticide.

Chemical Detox Procedure

1. See a clinical ecologist, who tests for chemicals, and find out if you do have a chemical addiction and exactly what is causing your symptoms. (If you need help locating a doctor in your area, see Appendix L for the address of the Society for Clinical Ecology.)

2. Avoid the offending chemical as much and as quickly as you can. If you cannot escape the substance or find a substitute for it, improve ventilation with a charcoal air purifier or by opening windows, so that fresh air continually circulates around you. Take frequent breaks outside. If the streets reek of exhaust gases and industrial fumes, try deep breathing on the roof of your building.

3. Prepare for possible withdrawal symptoms when you reduce exposure. These symptoms, which range from headaches, stomach and muscle cramps to confusion, behavior changes, and depression, might last anywhere from one to four days. Recognize these symptoms for what they are and do not panic.

4. Try not to take medication. In the unlikely event that your symptoms become severe, call your doctor.

5. Increase your daily vitamin C intake 500 to 1,000 milligrams. In 1971, two Utah State University scientists, D. J. Wagstaff and J. C. Street, provided evidence that ascorbic acid increases the effectiveness of the specific liver enzymes that detoxify pesticides and other chemicals.

6. Be sure to get adequate food and nutrition at this time so that you maintain your normal weight. Dieting forces the breakdown of fat deposits, releases stored chemicals into the bloodstream, and adds to the toxic load, possibly causing or intensifying your symptoms. Diet later when you can do it gradually and your body can handle the detox processes with no strain.

7. Try to lower your fat consumption and take special care to avoid red meats. Beef fat usually contains pesticides from the grains used to fatten the cattle, along with antibiotics and synthetic hormones, and your body has enough work to do.

8. Once your withdrawal ends, your detoxified condition is relatively easy to maintain. As long as the substance is avoided, backsliding into addiction is uncommon.

9. Be forewarned that after withdrawal, exposure to even small amounts may give you an unusually strong reaction.

10. If at any time in the future you must come into contact with the offending chemical, wear protective clothing

and use protective equipment. Clean your garments after each exposure, and if you have to take them home with you, carry them in a sealed bag.

Cocaine

Sylvia D. grew up in a close family and at one time wanted to be a rabbi. Her skills led her in more lucrative directions, however, and at 33, she became a $50,000 executive with a giant oil corporation. Two years into the job, she was also into a costly cocaine habit. The once dynamic, motivated woman changed into an irritable, depressed paranoic, furtively snorting coke at her desk from a hollowed-out pen and refusing phone calls from her family.

The turning point was meeting a former lover on the street. He seemed so shocked at her appearance, she lied and said she had just had major surgery. At home, she stared into her mirror, cried, then flushed away $300 worth of cocaine and phoned a medical detox center listed in the phone book yellow pages. Two months later, she was physically clean, mentally sobered, and back with her family.

Despite rumors to the contrary, cocaine is extremely addictive, afflicting many users with an urge they cannot control. If you have been using the drug six months or less, you should be able to break the habit by yourself. Research reported by *California Magazine,* November 1983, indicates that fewer than 1 percent of people who try cocaine will develop problems that require professional care. According to reporter Bob Roe, "Users stand a 99 percent chance of never having to choose among the myriad treatments offered today for addiction."

Cocaine Detox Procedure

1. Throw away all cocaine and all paraphernalia associated with its use.

2. Sever connections with your suppliers. Tell them you are addicted; most will not deal with anyone whose lack of control or death from an overdose could alert the police.

3. Tell family and friends that you are quitting and need their support.

4. Anticipate symptoms and keep in mind that the first two weeks are the hardest. Cocaine withdrawal is characterized by depression, anxiety, and lethargy, along with strong cravings for the drug. Heavy users may experience muscle aches and twitching, insomnia, nasal tissue irritation, and occasionally, delusions and hallucinations. All these feelings usually dissipate in a week. Taking tranquilizers to come down from cocaine can and often does lead to Valium or barbiturate dependency. Medication should not be needed, as cocaine users rarely suffer the more serious physical symptoms associated with alcohol or barbiturate withdrawal.

5. If you feel you require outside help, skip to step 12. If you want to try to quit on your own (help is always available), heed the words of Dr. Richard Louis Miller, founder of Cokenders, a residential program in Wilbur Hot Springs, California: "So you're going through pain with your withdrawal, and you've probably been taught to cover up your pain or take drugs to get rid of it. Make a major step toward personal freedom and do nothing but experience your feelings. You won't die, you'll just be uncomfortable.

"Close your eyes and look at your feelings of discomfort. Try to locate the spot in your mind. Point to it with your finger, then draw a circle around the feeling. Is this the feeling that triggers your desire for cocaine? What color is it? What is its shape?" Locating and focusing on the feeling, according to Dr. Miller, will help you control and change it.

6. You may feel other emotions: a sense of weakness, inferiority, guilt, and hopelessness, along with the fear that you may not be able to stay clean. Shake your negative image and rebuild your self-respect and confidence with

positive activities. Do the things you do best and enjoy the most.

7. For a detailed program of diet and exercise instructions, as well as a complete list of vitamin, mineral, and herbal supplements that may aid the process, read *Getting Off Cocaine* by Michael Weiner, Ph.D. The May 1984 edition of *American Health* magazine reported new experiments supporting Weiner's theory that the amino acid *tyrosine,* a protein building block usually found in meat and wheat, can help overcome cocaine depression.

8. Do heavy exercise for at least twenty to thirty minutes a day. The endorphins it produces reduce cocaine hunger and the anxiety that often accompanies withdrawal.

9. Keep busy; fill your spare hours with new interests. Make lists of things you have always wanted to do, and if they seem expensive, think about what your coke habit costs.

10. Avoid places where you might be tempted and persons who might tempt you. Look for a new set of friends.

11. Make love with your usual partner. David Britt, author of *The All-American Cocaine Story,* suggests, "Grab your mate and do what comes naturally. Making love can be a great diversion, an emotional release that may well get you temporarily past the need for cocaine."

12. If you wish professional help at any time, consult your family doctor and see what kind of therapy he recommends. Also consider talking to friends who have quit and find out how they did it. Call hospitals and detox clinics listed in the phone book yellow pages and find out what services they offer. Your choice of a center should depend upon whether you require live-in or live-out care. (See steps 14 and 15.) Look into support groups of nonusers—people who meet on a regular basis for mutual reinforcement. (Appendix L has a list of resources, including two organizations that tell you how to start your own support group.)

13. Call 800-COCAINE, a free, twenty-four-hour hot line sponsored by the Fair Oaks Hospital in Summit, New Jersey, where trained counselors will offer advice. According to their brochure: "800-COCAINE refers callers from any of the fifty states to a nationwide network of treatment facilities. One hundred of these state centers offer inpatient services; approximately 250 are physician's offices and clinics. Referrals to this network are accomplished through a computer-supported system, so callers get immediate information on the nearest center."

If money is a problem, call your local or state information center and ask for the nearest substance-abuse agency. Every state has one; its purpose is to coordinate available treatment centers and refer you accordingly. These offices are not connected with law enforcement and will not report your drug use. Most major cities have excellent, medically staffed free clinics.

Treatment regimens vary widely. Medical hospitals usually follow well-structured routines that offer Valium, Clonidine (an antihypertensive), or even small amounts of alcohol to help patients "come down" from coke. Other centers, such as Dr. Miller's Cokenders, avoid medication in favor of nutritional guidance, group and environmental healing, massage, art, acupuncture, exercise, meditation, and hydrotherapy (mineral springs baths).

14. If your cocaine habit is not excessive, and you are still reasonably rational and in control of your life, try *nonresidential* care. Dr. David E. Smith, founder of the Haight-Ashbury Free Medical Clinic in San Francisco, believes that the best treatment is "a combination of individual and group therapy provided in an outpatient setting. The process requires substantial education of the client on addictive disease and on cocaine use specifically. In many cases, recovery is enhanced by family therapy."

Some therapists use a controversial method called the contingency contract—a form of self-blackmail. The patient writes out a $5,000 check to a cause or a politician he hates, or types a letter to his employer admitting he is a drug addict; he also agrees to regular urine monitoring. If any samples are cocaine positive, the therapist is instructed to mail the check or letter.

In a 1981 study sponsored by the National Institute on Drug Abuse, thirty-one of thirty-two patients on contingency contracts produced cocaine-free urine for three to six months. The thirty-second patient voluntarily told his employer about his problem—this was the contingency—and entered a residential program. He kept his job and presently remains abstinent.

15. Dr. Ronald J. Catanzaro, chief medical officer of the Palm Beach Institute in Florida, feels that any or all of the following symptoms are serious enough to warrant *residential* treatment:

One's life becomes centered around the use of cocaine;

The user's life-style deteriorates, including losing contact with family members;

Incurs debt because of buying cocaine;

Feels emotionally unstable, depressed, and paranoid;

Is undernourished and appears chronically ill.

If you do choose a residential program, be sure it offers two to three days of medically supervised detox treatment, therapy that encourages you to confront the personality traits and weaknesses which led to your drug dependence, education about your illness, and meetings with family members. Optional services may include commitment to a support group such as Narcotics Anonymous or Cocaine Anonymous, and spiritual guidance.

16. When you feel the urge to backslide, call a drug-free friend and commit to an immediate activity—dinner, a movie, a set of tennis, or just taking a ride together.

17. Leave any location or party where coke is being

used *immediately.* As with all drug addictions, abstinence is crucial. Dr. Miller does not tell clients that they can never again have cocaine. "Never is a long time," he says, "but I do tell them that if they use coke again in some social way, the chances of immediate addiction are very high."

Dr. David Smith is more adamant: "In no sense can a person return to controlled cocaine use. Silence is the enemy of recovery. Open discussion of alternatives is vital to prevent first use and relapse. Recovery means *no use* and can be a very positive, life-enhancing process."

18. If you do slip, analyze your mistake and learn from it. Call a friend in your support group right away. Schedule a full calendar of activities for the times when you might backslide again, involving commitments to the people you most care about.

19. Maintain abstinence with support groups, aftercare from your therapy program, and continued exercise—probably the single most productive activity you can do. It keeps you in touch with your body and helps you develop respect for this marvelous machine.

Food

Waking up was a gloomy experience for Camilla P., yet the world would brighten considerably after her morning toast and coffee. Without knowing it, Camilla was hooked on wheat. Her sensitivity to this food caused her to depend on it to keep her metabolism in its altered state. When the wheat was removed, the metabolic functions previously acting at a certain level had to be readjusted, causing the sensations of withdrawal.

In her office at the bank, Camilla would usually eat a diet lunch of fruit and cottage cheese. By mid-afternoon, she would feel unaccountably depressed, grouchy, and angry— the beginning of withdrawal symptoms. Two cookies gave her a quick fix and picked her up. Rolls at dinner fulfilled the same function; so did crackers at bedtime.

A steady weight gain led Camilla to see Dr. Saifer, who noted her dependence on bakery products and suggested a four-day elimination diet, consisting mostly of foods she rarely ate. Camilla underwent forty-eight unpleasant hours of jittery nerves, stomach cramps, and urinary frequency. By the third day, her once bloated body was eight pounds lighter, water retention having been part of her altered metabolism. Her head was clear, and she felt no urge to attack the cookie jar. After six months of abstention, Camilla found she could eat small amounts of wheat every four days and not suffer a reaction.

Learn to recognize the signs of food-sensitivity addiction. Some people are simply voracious eaters; they gorge themselves regularly on pasta, pizza, and pastry because they like the taste. When denied food, they may be frustrated or unhappy, but they are not physically addicted and will not suffer symptoms. Most likely they will feel less bloated, more energetic, and physically healthier if they cut down their food consumption.

The reverse is true of those with a food-sensitivity. When their particular food—be it wheat, corn, eggs, beef, or whatever—is unavailable they become physically and mentally ill.

Answering yes to any of the following indicates that you may have a food addiction:

Do you often find it impossible to resist eating a certain food?

Do you ever feel desperate—that you would go anywhere or pay any price for that food?

Do you feel extremely weak, irritable, and anxious until you satisfy your craving?

Do you suddenly feel animated and energetic when you eat the food?

Food addiction leads to a repeated cycle of getting a fix, experiencing the first pangs of withdrawal, then taking another fix. If no fix is available, you may experience such symptoms as chills and trembling, sweating, anxiety, confusion, headaches, itching, and even vomiting and diarrhea, much like a heroin addict's experience of withdrawing cold turkey. Some foods, such as milk and wheat, contain opioids—morphine-like substances—and are therefore more likely to be addictants.

In the words of Dr. Theron Randolph, "A person who demands his meals or drinks on time in order to avoid becoming sick or generally miserable . . . gives himself away as a probable food addict."

Food-Sensitivity Detox Procedure

1. Determine what is (or are) the food offender(s). Do this by challenge-testing yourself. The procedure involves eliminating each suspected food, one at a time, then reintroducing it to be certain it is a culprit.

Shirley L., an art gallery curator, suffered symptoms for several years before she began to suspect her chronic fatigue had something to do with milk. She decided to challenge-test herself by eliminating milk and milk products from her diet for a week, but the results were disappointing. She soon began reading labels, however, and was amazed to find that milk or milk derivatives such as lactose and caseinate were in many foods she had been eating regularly, thinking they were milk-free.

When she finally eliminated *all* milk products from her diet, her energy began to return in three days. She challenged herself with a tablespoon of milk and almost immediately felt like "a wet dishmop" again, proving that she had no tolerance for milk.

2. Stop eating the offending food or foods and all products that contain them. Ask your doctor for a special diet for two to four days while your body adjusts to its newly nontoxic state.

3. If your doctor is a disbeliever in food addictions, you may want to consult a clinical ecologist who understands the problem and will offer psychological support through a possibly uncomfortable withdrawal. The worst of your symptoms should abate within forty-eight hours.

4. Institute the rotation diet described in Chapter 4, and make this a permanent way of eating. Your food addiction problems should not recur, except, possibly, in one very troublesome area: the sweet tooth.

SUGAR

In his last year of college, Mark E. realized that his sugar consumption, in the form of eight to twelve candy bars a day, was excessive, helping neither his weight problem nor his health. A young lady he was seeing presented him with a copy of the book, *Sugar Blues* by William Dufty; he read it and could no longer ignore his problem. He knew he had to break his addiction—but how?

Unwisely, he took a friend's advice to go on a rice-and-bananas diet, omitting sugar suddenly and completely. On the second sugar-free day, his body "crashed," and he passed out in the classroom, later awakening in a hospital bed.

Dr. Saifer was called in to treat Mark and immediately recognized the signs of sugar addiction. She explained to him that although blood sugar levels are boosted by a sucrose "rush," they drop within the hour, after the sugar has been metabolized, and leave the person feeling droopy and irritable. To regain energy and vitality, the person eats more sugar, gets a new high, and the cycle continues, creating peaks and valleys of blood sugar concentration that produce a true physiological dependence.

With his doctor's approval, Mark entered a private detox center, and after a week of education, therapy, and bouts of depression, anxiety, and cravings, Mark was able to go home and throw away all his candy bars. His new diet—heavy on fresh fruits, vegetables, fish, fowl, and nuts, and low on animal protein—keeps his weight controlled and his energy levels stable.

Sugar masquerades as a harmless food, but is actually an insidious drug that creeps into everything from toothpaste to Teriyaki sauce. If you're given to joking about your sweet tooth, as many of us are, consider the possibility that you may be unknowingly addicted. Sugar yearnings that appear at unusual times may signal a serious problem which requires the help of a clinical ecologist.

Some women, just before their periods, develop an intense craving for sugar (usually chocolate bars) as part of the "premenstrual syndrome." If it lasts only a day or two, it does not warrant serious attention; if it lasts for a week or more, it needs a doctor's care.

Sugar craving can also develop when sensitive people are exposed to certain chemicals. One such case was Maribeth L., who kept hard candy in her old jalopy because she knew her sweet tooth would act up as soon as she started driving. Since Maribeth had many chemical sensitivities, Dr. Saifer guessed that her car candy hunger was another manifestation of this illness. Sure enough, when Maribeth repaired the exhaust leaks in her car, and was no longer inhaling the gas to which she was sensitive, she was able to drive without craving sugar.

Finally, a number of people have the yeast problem so well described by Dr. William Crook in his 1983 book *The Yeast Connection*. This yeast or fungus, *Candida albicans*, normally inhabits the bowels of human beings from shortly after birth. The use of oral contraceptives, antibiotics, or cortisone can increase the yeast population so that it becomes an allergic overload for sensitive persons, and causes the illness known as candidiasis, a symptom of which is sugar craving. Doctors often think of this possible problem when people report sugar addiction. Once the yeast are killed by the appropriate medication, nystatin, the sugar craving goes away.

It is important to remember that sugar addiction can have a number of causes, many of which are potentially serious.

Before you begin the sugar detox procedure, form the habit of reading food labels carefully. Ingredients are listed in order of amount used, so if sugar, honey, corn syrup, or dextrose are among the top three ingredients, they are present in large amounts. Don't be misled: a barbecue sauce might list sugar second and corn syrup sixth, but if added together, sugar would be the main ingredient. Another possibility is that sugar may be "camouflaged" under a different name. Fructose, glucose, lactose, maltose, sucrose, molasses, maple syrup, and sorghum are all sugars.

Sugar Detox Procedure

1. Keep a diary for a week in which you write down how many sugared products you eat and drink in a day. While you may not be able to figure out exactly how much sugar you are eating in a slice of bread, for instance, if sucrose or dextrose are high on the list of ingredients, the sugar content is greater than you think. Most doctors consider half a cup of sugar a day to be excessive.

2. If your sugar intake is large, cut back gradually. A cold-turkey withdrawal may shock your system. Even if your sugar intake is only a few teaspoonfuls a day, reduce it. Treat sugar as a delicate condiment rather than a staple.

3. Eat no sweets between meals. Sugar on an empty stomach starts you on the rush-crash cycle. Sugar eaten after a meal is slower to enter the bloodstream, and its effects are lessened. Instead, nibble on nuts, seeds, unsweetened granola, unbuttered popcorn, and raw fruit and vegetables. Have supplies handy in home, car, and office. "The trick to avoiding sugar pangs is to keep the blood sugar fairly even," says author-pediatrician Dr. Lendon Smith. "If you nibble every two to three hours on good food, you'll eat less at meals and you'll crave less."

4. Avoid red meats. *Sugar Blues* author William Dufty writes that, "In kicking sugar, the most helpful hint I can give you is the one that worked for me. Kick red meat. Just switching to fish or fowl reduces your desire for a sweet concoction at the end of a meal—makes it easier to settle for natural fruit or for no dessert at all."

5. Take B complex vitamins as suggested in the General Detox Program in Chapter 4. They help assuage sugar hunger, according to Dr. Smith. Whole grains, wheat germ, fresh vegetables, bananas, nuts, and fish are good natural sources.

6. Consider taking 15 milligrams of elemental zinc daily. Tests reported in *Prevention* magazine, September 1983, show that zinc enhances taste sensitivity. Subjects who took zinc supplements for fifteen weeks found that they could get by on less sugar. Check with your doctor first.

7. Banish sugar. By now you should be ready to clear your house of all sweets, including corn syrup and molasses. Keep one small jar of raw honey or pure maple syrup for special treats.

8. Banish sweeteners. "They are chemicals and the liver has to detoxify them," says Dr. Smith. "Besides, anything that tastes sweet promotes the idea that everything must taste sweet." Reeducate your taste buds.

9. Splurge once a week on a sweet, if you so desire. Unlike alcohol and most other substances, moderate sugar use does not lead right back to addiction. You will probably find that the long-anticipated dessert does not taste half as good as expected, and eventually, all types of sweets will turn you off.

Marijuana

One evening after his Boy Scout meeting, twelve-year-old Paul P. walked home with Jeff, the son of a neighbor.

Jeff invited Paul to his garage and asked if he wanted to try marijuana. Soon, Paul and several of his friends were buying it from Jeff and smoking it regularly. Paul's grades began to decline, and he often came home spacey, talkative, and disoriented. His parents discovered why, but their efforts to reason with him, bribe, and punish him all failed.

At age sixteen, Paul became infatuated with a young gymnast who told him he was poisoning his body. He tried to stop cold turkey, but found himself uncomfortably nervous, took a few drags to calm down, and was hooked again. The trauma of losing his girlfriend made him recognize his addiction; fortunately, he was able to talk to his parents. They agreed that the best solution was a change of high school. His withdrawal, without the peer-group pressure, was surprisingly easy. Paul developed a new set of non-pot smoking friends and carries a picture of his ex-girlfriend in his wallet, as a deterrent to backsliding.

The happy part of Paul's story is that he stopped. Many young people either continue smoking pot, into their forties and beyond, or go on to stronger drugs.

Marijuana Detox Procedure

1. Quit immediately. Flush all marijuana down the toilet, break your paraphernalia, and throw away the pieces.

2. Tell your dealer you are stopping. You owe no explanation to anyone, but if friends persist, say "health problems."

3. Tell whatever family members know of your habit that you are quitting. Solicit their extra love, attention, and support.

4. Expect some symptoms. The commonly voiced assurances that marijuana is neither addictive nor harmful are untrue.

A 1982 study, led by Dr. David W. Schnare of the non-profit Foundation for Advancements in Science and Education, found that THC, the psychoactive ingredient in marijuana, accumulates and stores in the fatty tissues of even casual smokers. Under conditions of stress, weight loss, exercise, heavy perspiring, or just stopping usage, THC moves from the fat into the bloodstream. THC release can cause flashbacks as well as the classic symptoms of marijuana withdrawal: hyperactivity, insomnia, decreased appetite, and anxiety.

5. Psychological habituation to marijuana is less than cocaine dependence and rarely requires medical treatment. If, however, you know or suspect that your marijuana has been laced with the dangerous chemical PCP or other adulterants, do seek medical care as your withdrawal may produce complications. (See Appendix L for list of drug-abuse resources.)

6. You may feel hyperactive for one to three days. Use that surplus energy to clean closets, go to the library, visit friends, join a gym, or do all the things you have been putting off. Read, watch television, exercise, make love. Plan your days so you end with the most enjoyable tasks.

7. Take no medication for insomnia. Your body is readjusting and you should be sleeping better than ever in several days or a week.

8. Expect a slight weight loss. Your appetite will return in a week or less.

9. Look up the alternatives to smoking in Chapter 4, and read the program to reduce cigarette smoking dependency, which comes later in this chapter.

10. Avoid persons and situations where you might be tempted. Seek new friends and activities. Immediately leave any place where pot is used.

11. Reward yourself. Spend your marijuana money on the best hairdresser in town, seats at the theater, a Chinese dinner—anything to ease the emotional pangs and encourage you to maintain your healthy, detoxified body.

Prescription and Nonprescription Medication

Medical drugs help many more people than they harm, but they all have side effects, not the least of which is that they are often addictive. If you are a regular user of any medication, the detox process involves taking steadily decreasing doses of the substance, or substituting a less toxic, less addictive drug and then steadily decreasing its dosage.

A West Coast school administrator, Angela R., had frequent nausea and "spacey" feelings while driving to and from work. Her doctor could find no physical cause. He diagnosed "job stress" and prescribed 5 milligrams of a tranquilizer, three times a day, assuring her that the drug was harmless and she could stop taking it at any time. He also warned that it would make her too sleepy to drive a car.

Angela's symptoms diminished almost immediately, and three months later, still on the tranquilizer, she decided to try driving. When her nausea suddenly reappeared, she realized that the tranquilizer was not helping and promptly quit the medication. Within twelve hours, she was feeling terrible—depressed for no reason, nervous, irritable, and unable to eat or sleep. In short, she was addicted.

Angela was appalled at the intensity of her cravings and suffered several days of discomfort. Thanks to a loving family, her mood lifted in a week, and her symptoms subsided. A newspaper article about Dr. Saifer opened her eyes to a possibility she had not considered; she visited the office, was tested, and proved to be highly allergic to auto exhaust.

Dr. Saifer suggested that Angela install an activated charcoal air filter in her car, drive at times of minimum traffic, and use every possible means to reduce exposure. Angela's symptoms cleared again, her health returned, and her only regret is that her body was saturated for so long, with unnecessary medication.

Medication Detox Procedure

1. Learn all you can about the drug you are taking. Ask your doctor or druggist, check a medical reference, or write to the pharmaceutical firm that makes it. You should determine how addictive it is; what the likely withdrawal symptoms are, if any; how long the symptoms last; and means to ease withdrawal.

2. Suspect addiction if:

You feel you need the drug to function normally.

You are tempted to take a higher dose to have the same effect.

Missing even one dose makes you feel sad, sick, or nervous, and experience craving.

You are indifferent to the cost.

You continue to use it in spite of negative side effects or consequences.

You or members of your family have a history of substance abuse.

3. Never stop using a prescription drug unless you have strong reason to believe it is harming you. Call your doctor immediately. He may have reasons for keeping you on the medication that he has not fully explained. Some physicians may not support your need to quit and will assure you that the medication is safe for as long as you want to take it. This is not always in your best interest; pharmacological treatment is a valid medical procedure mainly as a temporary measure or when there is no alternative.

4. If your doctor denies the possibility of addiction and

you feel otherwise, seek a second or third medical opinion from doctors with different orientations and then make your decision. If you decide to continue with the medication, at least you will know why you are doing so.

5. If you choose to stop the medication, ask your doctor what to expect and what to do if your reaction is severe. Plan to be busy on the day you stop or begin to taper down, and not to be alone. Withdrawal symptoms can range from minor skin rashes, weight loss, headaches, stomachaches, insomnia, nausea, dizziness, and disorientation to such major symptoms as heart trouble, grand mal seizures, and temporary psychosis, including paranoia, hallucinations, and loss of touch with reality.

6. If the dependency is minor, your doctor may agree to help you taper off. If you need a new supply of medication, have him write "no refill" on the prescription. Breaking the sleeping pill habit, for instance, can often be done at home and, if there are no physical or emotional complications, with minimal medical guidance.

7. Keep accurate records of your daily dosages, and whatever symptoms appear, if any, with slightly lowered amounts. Even with gradual withdrawal, you may experience sadness, worry, anger, sleep disturbances, and reduced appetite. These will diminish and disappear in several weeks; if your symptoms become severe at any time, call your doctor immediately.

8. Whether your dependency is minor or major, your doctor should warn you about the possibility of a strong reaction and may suggest using substitute drugs to ease the withdrawal. These can be administered either at outpatient or inpatient facilities and require close medical supervision.

9. Look into hospital and private detox units, community centers and clinics, and other resources. Your doctor should be able to select one best suited to your personality,

your particular medication, and your finances. The local medical association may know of private doctors in your area who specialize in drug toxicity.

If you are pregnant, or suffer chronic mental or physical disorders, residential treatment is recommended. Some medical insurance policies will cover it.

10. Be wary of alternative therapies suggested by well-meaning friends. Hypnosis, for instance, is virtually useless in cases of medication withdrawal; it will not alleviate symptoms and rarely helps reinforce willpower.

11. Once your withdrawal symptoms have abated, you may have some success with acupuncture, biofeedback, and behavioral modification techniques to help you form new habits that reduce your need for medication. These therapies produce best results when combined with counseling and family or group support.

12. To prevent future addiction, explore with your doctor or therapist the reasons for initial use of the drug. Also consider the possibility that your depression, hyperactivity, or whatever prompted you to medicate might be a toxic reaction to something in your environment. The worst possible treatment is to cover one chemical reaction with another.

13. Be alert; become an informed consumer. To help maintain your medication-free state, remember that the major drug companies spend a fortune each year to make their products known and appealing to you and your doctor. Lifetime, the cable health channel, recently reported that the drug industry in the United States will make a net profit of $4 billion in 1984! Do not place blind trust in any drug or any doctor. Take *no* medication until you have ascertained that it is absolutely essential to your health and well-being.

Nicotine

An interior designer, Earl W., smoked three packs of cigarettes a day and had tried to quit several dozen times without success. When a close smoking buddy died of lung cancer, he thought about quitting again. He even tossed his cigarettes into the wastebasket, only to retrieve them two hours later when the beginning withdrawal symptoms of irritation, confusion, and anxiety were too much to handle.

At a dinner party given by a client that evening, he was the only one at the table who smoked; his hostess expressed her displeasure by asking in a loud voice how he could enjoy his food with smoke-dulled taste buds. Because of his friend's death and his own frustration at not being able to stop, Earl was overly sensitive to the subject. He took offense, excused himself, and went home.

The hostess phoned the next morning and dismissed him from an extensive redecorating job. Horrified at the loss of a six-figure contract, he sent her an extravagant gift with apologies and promised never to smoke again if she would forgive and reconsider. She did so, supplying the motivation he needed. With the support of friends and the help of doctor-prescribed nicotine chewing gum, he was finally able to break the stranglehold.

Ninety-five percent of all smokers who make up their minds to stop, do so—"just like that." They may suffer physical and psychological withdrawal symptoms, but once they get through the first week they generally stay clean.

Swedish researcher Karl-Olov Fagerstrom has identified four types of smoking dependency:

Type 1 is the person who lights up in social situations. It gives him something to do with his hands and eases the tension of group gatherings. He smokes for the image.

Type 2 is the person whose habit centers around daily activities—coffee breaks, lunches, telephone calls—as well as social functions. His attachment is mainly conditioned and psychological.

Type 3 has a physiological or chemical dependency and can be as hooked on nicotine as an alcoholic is on his bottle.

This person lights a cigarette before breakfast in the morning, smokes all day long, and the last thing at night. He needs to maintain a certain plasma nicotine level to avoid the symptoms of withdrawal.

Type 4 has a chemical and an emotional dependency, smokes even more heavily than Type 3, and inhales more deeply.

Types 1 and 2 can usually stop smoking with an effort of will; their habit is not too hard to break. Type 3 is best helped by combining physical aids or alternatives with psychological support. Type 4 may need clinic or hospital care. According to researchers at the Stanford University Medical Center, only about 5 percent of smokers seek outside help in quitting.

Steps 1 through 19 of the Nicotine Detox Procedure are for those who think they can stop smoking independently. Professional help is outlined in steps 20 and 21.

Nicotine Detox Procedure

1. Make a firm decision to quit. Set a target date four weeks hence. Follow steps 2 through 8 during those four weeks.

2. Ask family and friends to reinforce your intention. Make a bet with someone that you will stop on your target date and stay clean for six months afterward. Invite a friend to quit with you.

3. Keep a journal with each day divided into hours, marking the exact time of every cigarette you smoke. Count the total each night before you go to bed.

4. Cut down usage. Postpone lighting each cigarette for five minutes the first day, six minutes the second day, and continue adding a minute a day. While waiting to light up, do something enjoyable, engrossing, and distracting. Watch the number of cigarettes you smoke decline.

5. Switch to a brand you dislike, smoke with the hand you don't normally use, puff only half a cigarette.

6. For oral gratification, drink liquids through a straw. See the list of alternatives in Chapter 4 and have some "props" on hand.

7. Try conditioning yourself to dislike tobacco. One aversion therapy is to leave ashtrays dirty for several days; if your family can stand it, the stale smell will help turn you off. Another technique is to practice rapid smoking. Inhale deeply every six seconds for five minutes, rest, then repeat the procedure until you can no longer inhale the smoke. You may feel some distress in the form of nausea or a headache. (This technique could strain the heart and should be medically approved.)

8. Put your cigarette money in a piggy bank to be opened after six months of success, and then spend it on a wild indulgence or a donation to the American Cancer Society.

9. On the day you quit, throw away all cigarettes. Remove ashtrays and matches. Follow steps 10 through 13.

10. Keep busy every minute. In nonworking hours, do things you enjoy with people who will be supportive.

11. Have your dentist clean your teeth and remove all nicotine stains.

12. Avoid situations where you used to smoke. If this is impossible, and someone questions you, simply say, "I quit." Do *not* say, "I'm trying to quit." Remember that society is now on the side of the nonsmoker.

13. Reward yourself. Buy a luxury item, plan a trip, take your family to an elegant dinner. You are only spending your cigarette money.

Once you quit, you may have no symptoms whatsoever; but if you do:

14. Expect the most distress the first three days; irritability, depression, anxiety, muscle cramps, headaches, and sleep disturbances are all possibilities. Craving for tobacco, the most frequent and difficult symptom, usually reaches its peak twelve to twenty-four hours after the last cigarette and tapers down over a seven-day period, but it may return and persist for up to eight weeks.

You may also experience bowel irregularity, sore gums or tongue, and temporary weight gain caused by fluid retention. According to the nonprofit group Action on Smoking and Health, "It is uncertain whether weight gain is caused by increased appetite, a changed metabolism, or both." The risks of carrying a few extra pounds, however, are far less than the risks of smoking. Your physical symptoms should all be over in a week; the psychological need for a lift, or something to do with your hands, may continue for months.

Amazingly, these withdrawal pangs are really symptoms of recovery; only six to nine hours after you stop smoking, the physical harm done starts to reverse itself. Your heart rate and blood pressure slow; at the same time, the carbon monoxide level in your blood drops rapidly, allowing the damaged cells in your heart and lungs to begin the rebuilding process.

To aid withdrawal:

15. Ear acupuncture may decrease physical symptoms.

16. Hypnosis generally does not help physical symptoms, but it can be useful in changing your self-image from smoker to nonsmoker or in handling the emotional aspects.

17. Cut down your intake of meat, fish, and seafood (except shrimp), eggs, and poultry, as these foods make acidic urine. Eat plenty of fruits (except plums, prunes, and cranberries), vegetables, nuts, and seeds, as they form alkaline urine. The newest theory, supported by a number of physicians, is that alkaline foods take longer to flush nicotine

out of the system; therefore, the nicotine stays longer in the body and decreases the intensity of the craving.

18. Take two tablets of Alka Seltzer Antacid Formula in gold (not blue) foil, dissolved in water as an alkalizing agent. Take one gram of vitamin C every three hours during the day and, if possible, at night. Drink large amounts of fluids.

19. Reinforce your willpower with visualization. Conjure up images of smoke-blackened lungs or of lying on an operating table or in the hospital's intensive care unit attached to various tubes and machines. Positive images work equally well; see yourself as healthy, active, running up stairs without gasping for breath, and enjoying the social and personal rewards of quitting.

If you have tried to quit on your own and failed, or if you simply feel you cannot handle the problem alone, then you need professional medical care:

20. Check with your doctor. When discussing a clinic, hospital, or private program, ask the questions listed in Chapter 4, along with:

Do you provide alternatives to smoking?

Do you deal with weight gain?

Is there a followup system to help resist temptation in a smoking environment?

21. See Appendix L for a list of resources and support groups.

If you relapse:

22. Watch for specific situations. Paradoxically, both moments of pleasure and moments of unhappiness can trigger a relapse. University of South Florida psychologist Saul Shiffman and Dr. Murray E. Jarvik at the University of California, Los Angeles have identified the four most dangerous threats to ex-smokers: drinking alcohol, especially in a group

of smokers; relaxing after a meal; feeling job pressures; experiencing boredom or depression.

23. Think back on how hard you have worked. Would you wipe out all that effort because of one little slip?

Four weeks later:

24. Your sense of taste has returned, your smoker's hack is gone, your digestive system is back to normal, and your head is clear—no more headaches, depression, or "short fuse" bursts of anger. You smell better. Nonsmokers no longer give you dirty looks or cover their faces. You feel alive, full of energy and vitality, and proud of accomplishing exactly what you set out to do.

6

Life After Detox: The Future of Nontoxic Living

In a sense, we are a lucky generation. The problems of environmental pollution and the proliferation of toxic substances in our lives are both tangible and controllable. Unlike death and taxes, we can avoid, stop, and even reverse the effects of many synthetic chemicals. In the last decade, public awareness has zoomed, and governmental agencies have forced higher standards for work conditions and waste disposal, food and drug purity, air quality, product safety, and much more.

The battle continues. In February 1984, more than 100 labor, environmental, and consumer groups announced plans for a national campaign to reduce exposure to toxic chemicals. John O'Connor, coordinator for the National Campaign Against Toxic Hazards, told a news conference that, "The invasion of toxic chemicals into our bodies is America's No. 1 hidden health problem. Each of us . . . carries detectable traces of cancer-causing chemicals."

On an individual level, we have learned how to detoxify our bodies and keep them that way. There is no doubt that we can live clean lives in a polluted world. We cannot, however, dissociate our fate from the fate of the earth. What we have learned about freeing our bodies from harmful substances must also apply to cleaning up the world. Researchers and scientists in every area of civilized life are continually seeking new nontoxic ways to preserve food, cure illness, produce products, and maintain our high living standards. By the year 2000, we should all be breathing cleaner air, eating unprocessed and uncontaminated foods, and drinking water that is fit to drink.

HOPES FOR TOMORROW

Recent scientific and technological advances in several areas have already begun to pave the way for a nontoxic future.

Air Pollution. The U.S. Agricultural Research Service now has the technology to strip burning coal of the sulfur and nitrogen that cause acid rain and to convert some of that residue to plant fertilizer. How to dispose of the remaining waste is temporarily unsolved.

Various plans are also under way to reduce car exhaust. Chrysler has bought the patent for a mechanical valve that sits beneath the carburetor, raises gas mileage, and cuts carbon monoxide emissions by almost 50 percent. It should be in production by 1986.

Ford Motor Company is testing methanol-fueled cars for possible use by 1987. A spokesman for the Celanese Chemical Company states that, "Compared to gasoline or diesel, methanol is clean-burning. Sulfur emissions are eliminated, nitrous oxides are reduced, and ozone formation is decreased."

Alcoholism. Harvard Medical School doctors have isolated a compound that is present in the blood of alcoholics when they drink, but absent in the blood of nonalcoholics. This has spurred researchers to look for biological—rather than psychological—causes of alcoholism. The disease may one day prove to be curable.

Asbestos. An engineering professor at the University of

California, Los Angeles has developed a cement-strengthening material that is both cheap and safe, and which can replace asbestos in construction work. The formula combines scrap metal, marble, and slate and produces a glass fiber as strong and alkali-resistant as asbestos, but the new material does not produce carcinogenic dust.

Birth Control. One method in the works is a nasal hormone spray that would have fewer side effects than the Pill. Doctors at the University of Texas Health Science Center at San Antonio report that this new hormone blocks the release of eggs from the ovaries and that toxicological studies show it to be "completely safe." The FDA is currently considering it.

Cosmetics. The Center for Alternatives to Animal Testing at Johns Hopkins University is perfecting a technique that will allow you to test your skin reaction to a toiletry or cosmetic before you buy it. It is known that inflammation causes the production of oxygen; therefore, the amount of oxygen released from a sample of skin cells challenged by a particular product will give a clear indication of whether or not you will react.

Drug Addiction. A promising new withdrawal aid comes in the form of a small black box that supplies NeuroElectric Therapy (NET). "The treatment," says rock star Peter Townshend, who claims it weaned him from a variety of pills and drugs, "is effective for booze, cigarettes, barbiturates, cocaine, marijuana, you name it. There's a different frequency that works best for each kind of addiction."

Invented by Southern California surgeon Dr. Margaret Patterson, the Walkman look-alike clips to the belt and has two wires that you attach behind the ears. "It's a method of rapid detoxification," she explains. "The electricity quickly cleanses the addict's system of drugs and restores the body to normal within ten days."

Exactly how it does this is uncertain, but the best guess is that NET transmits tiny electric signals that seem to har-

monize with natural brain rhythms. These stimulate the brain
to produce its own sedative chemicals that help the body to
heal itself.

The box is currently undergoing FDA tests and, if approved, should be available by 1985. Individual models will
cost $1,000 or more, and as its inventor warns, "We can get
people off whatever drug they're hooked to, but it's up to
them to fill the void. They've got to find a constructive substitute for the drugs that have dominated their lives."

Dump Sites. The EPA has approved an engineer's pro-
posal to clean up toxic soil by heating contaminated mud,
converting its toxic chemicals into gases, capturing them in
filters, and burning them. The engineer plans to build an
$8 million pilot plant on the shores of Illinois' Waukegan
Harbor. It will process fifty tons of sludge an hour.

Fertilizers. Scientists at California's International Plant
Research Institute near Silicon Valley claim they are isolating
genes, cloning cells, and creating hybrids of plants and foods
that are so hearty and super-nutritional that chemical fertilizers will soon be obsolete.

Food. Research is turning up evidence daily about the
valuable use of nutrition to protect the body from toxins. It
has been determined that liver and green beans are high in
vitamins A and B_6, and can increase resistance to aflatoxins—
poisons produced by molds in cereals and milk, and on corn
and peanuts. Carrots and peas contain zinc and vitamin C,
and can help neutralize pesticides and nitrosamines. Carrots,
spinach, broccoli and cantaloupe also contain carotene, which
converts to vitamin A in the body, and has been shown to
be anticarcinogenic in rats and mice, and possibly humans.

Vegetables in the cabbage family contain substances that
contribute to the production of glutathione, an antioxidant
claimed to have anticarcinogenic properties. Milk, cheese,
eggs, and liver supply vitamin B_{12}, which reduces the effects
of the cyanide in cigarette smoke; and a low-fat diet decreases the odds of getting cancer from the potent hormone

diethylstilbestrol (DES), banned in 1979 in animal feed, but still found in tiny amounts in some poultry and livestock.

Herbicides. The EPA recently cleared its first natural herbicide, a fungus that kills northern joint-vetch, a weed that smothers rice. Research continues.

Medication. Drug-delivery systems are being studied at the University of Wisconsin. New techniques include targeting drugs to their specific organs in the body and directly implanting a drug near the target organ, thus reducing both the dosage needed and the toxic side effects.

The same team is developing one-a-day pills that remain in the gastrointestinal tract for twenty-four hours (the old limit was twelve hours) and provide a continuous low level of medication throughout the day instead of a high level all at once. This will also reduce side effects and lessen strain on the liver and kidneys.

Harvard Medical School physiologist John R. Pappenheimer has identified a chemical he calls Factor S.—the natural sleep-causing agent in human beings. One-trillionth of a gram of Factor S. injected into animals causes five to twelve hours of deep sleep. Pappenheimer and his team expect to produce a synthetic version in the form of a nontoxic pill that will make barbiturates and other addictive medication unnecessary.

The theory behind Factor S. is best explained by Dr. Seymour Rosenblatt and Reynolds Dodson, authors of *Beyond Valium*: "We shall emerge into a drug-free society (because) . . . the substances produced by our biochemists will match exactly the substances our bodies produce naturally." Medication of the future, they imply, will be free of all side effects.

Oil Spills. A Swiss company, Bregoil Sponge International, has invented a spongelike material that can be sprinkled over oil and certain chemical spills to soak up the substance and be easily scooped up. It works on land or water and is biodegradable. The wood-based product is currently available in Europe and is being test-marketed for sale in the United States.

Not to be outdone, Elf Aquitaine, the French state oil company, announced that they have developed a substance that helps particular oil-eating microorganisms multiply. The Microbes are attracted to the hydrocarbons in oil and are then used to "eat" oil spills.

Pesticides. Botanists at the University of California, Irvine have identified the chemicals that plants use to defend themselves from insects and are experimenting with "natural pesticides." They predict that fungi and bacteria-based pesticides will become a billion-dollar business by 1990.

Bio-Systems Research, Inc., in Salida, Colorado is planning to market a derivative of a tree chemical that deters boll weevils from feasting on cotton plants, and researchers at the University of California, Berkeley have isolated the active ingredient of an East African medicinal tree that is free of insects. They are testing the substance and hope to make it commercially available.

Pest management through sterilization of the males is being studied, according to a scientist at the U.S. Agricultural Research Service. He reports that the concept is not new, but advanced techniques for sterilization through breeding and irradiation are "revolutionary."

In February 1984, the FDA proposed regulations to allow the use of low-dose radiation to kill insects and bacteria on fruits and vegetables. While the FDA is convinced that the irradiated food will be as safe to eat as food cooked in microwave ovens, other scientists have voiced concerns, and the process is being reviewed.

A surely heartless method involves using an airborne sex pheromone—an odorous body secretion that an individual perceives as being from another of the same species—to lure male moths into mating with a different species of female moths. Because the insects are physically incompatible, they lock together and die within hours.

Toxic Shock Syndrome (TSS). The University of Wisconsin

is perfecting a simple blood test that will determine whether a woman is susceptible to this disease. A positive response would indicate that a woman should not use tampons, which have been linked with the bacterial growth that leads to TSS. She would also be warned to look for early signs, such as sudden high fever, vomiting, dizziness, and diarrhea.

Vitamins. A team of researchers from Sydney University recently discovered that an Australian fruit, the terminalia berry, has between 2,300 and 3,150 milligrams of vitamin C per 100 grams of edible fruit. By comparison, the same amount of orange contains only 50 milligrams of ascorbic acid. Nutrition researchers hope to find or possibly breed many more high-vitamin foods, so that we can ultimately get all our nutrients—even megadoses if desired—from our diet.

Wastes. Michigan State University professors have found a strain of bacteria that breaks down the potent chemicals in toxic waste and excretes them as harmless substances. Researchers hope to use genetic engineering techniques to breed a "superbug" who will dine on dioxin.

Water. Massachusetts Institute of Technology Professor Alexander Kibanov has developed an unlikely formula for detoxifying waste water. He mixes fresh horseradish, which contains the enzyme peroxidase, with hydrogen peroxide to form a compound that causes water toxins, such as PCBs, to solidify. They can then be removed as sediment or burned as fuel. His recipe awaits a patent.

Many fish have innate chemical detectors and can immediately sense the presence, even in minute quantities, of herbicide, pesticide, or industrial waste in their water. French scientist Jean-Louis Huvé has developed a method of implanting electrodes in a trout's brain that are coupled to a tiny transmitter. At the first "sniff" of pollution, the brain sends a warning to the computer. This delicate monitor could alert humans to take action before the pollution becomes a problem.

A four-year survey released in 1983 by the United Nations Environment Program found that despite the pollution of shorelines, the earth's 328 million cubic miles of oceans have become cleaner in the past decade. The 100 scientists who took part attributed the improvement to decreased use of DDT, PCBs, lead and other metals and chemicals as a result of worldwide cleanup efforts.

X Rays. The new Nuclear Magnetic Resonance machines use radio waves, magnetic fields, and computers to produce images of the body. Unlike X rays, they do not cause dangerous radiation effects. Hospitals that can afford this machine's $1.5-million price tag are now using the units with great success.

Many other techniques are rapidly making the standard X ray obsolete. Among them: ultrasound, which uses sound waves to make "maps" of the body; microdose digital radiography, a highly sensitive, computerized X ray which is virtually noninvasive; and the now familiar computerized axial tomography (CAT) scan that yields cross sections of the body.

Exciting as these new discoveries seem, they do not shift responsibility for a clean world onto the shoulders of scientific researchers. That awesome task still falls to you. One of the best ways to make a contribution is with your own body and your own personal environment.

MAINTAINING YOUR HEALTH

Before we consider the ways in which you can maintain the standards of health set forth in the General Detox Program, let us review your specific accomplishments.

You cleaned up your home and work environments by removing as many pesticides, chemicals, and toxic substances as possible. You threw out all products and cosmetics with warning labels, and you developed new nontoxic habits—such as brushing your teeth with baking soda, buying natural fiber clothes without chemical odors, and drinking filtered water.

You started a diet high in fruits, vegetables, whole grains, fish, and poultry, and low in organ meats and processed foods. You added vitamin supplements and learned to rotate the kinds of foods you ate.

You set aside a daily half hour for active exercise that stimulated your heart and lungs, and you were careful to breathe fresh, clean air.

You practiced a technique—be it meditation, self-hypnosis, or simply resting—to slow down every day, relax, and refresh your perspective.

With or without professional help, you were able to end your food or chemical addictions; stop using unnecessary medication; give up alcohol, caffeine, cigarettes, cocaine, or any combination of those drugs; and feel your body finally breaking free of its chemical captors.

Now the challenge continues. The longer you maintain your detox regimen, the stronger your liver, kidneys, and other excretory organs will become, and the more efficiently they will be able to metabolize the toxins you cannot avoid.

Although improvements are under way in many areas of life, most will take years to become realities. In the meantime, polluters are still polluting, illicit drugs are easily available, legal drugs proliferate at social gatherings, and medical drugs are as close as the nearest pharmacy. It seems that no matter where you go or what you do, someone will offer you a well-intentioned poison. How do you develop the strength to refuse it?

The answer is easy. If someone offered you rat killer, would you eat it? The active ingredients in this and in most insecticides are nicotine and arsenic, both of which are also in cigarettes and other "harmless" stimulants. You now know what these toxins can do to your body. Keep this knowledge foremost in your mind and remember that good health does not just happen; it must be earned, and in certain cases, fought for. When someone offers you a social drug, decline with grace and firmness. If the person feels rejected, he is right. You are not only rejecting his poison, you are rejecting his self-destructive philosophy.

The toughest adjustment will begin about three months after you put this book on a shelf. The immediate high that comes with the satisfaction of having cleansed your body or overcome an addiction is not permanent; nor are the social rewards that friends and family supply. When the glow diminishes, the temptation to revert to old habits will increase. You will be bombarded with the "why not?" and the "life-is-too-short" philosophies. The truth is that life is too short to be sick; and life is too short to spend the last half paying for the abuses of the first half.

People often speak of the "quality of life," a term which traditionally implied consuming—consuming television sets, food processors, and the latest models of fashion, furnishings, and fast cars. Today the emphasis is changing. We are discovering that the latest models of fashion, furnishings, and fast cars are literally toxic, and that the quality of life has little to do with possessions; instead, it means enjoying good health, nurturing a loving relationship, and spending your daytime hours at tasks that bring satisfaction. Your chances of achieving these goals are in direct proportion to your ability to avoid toxic substances and stay healthy. Who ever heard of a sick winner?

In the words of Joseph Califano, former Secretary of Health, Education and Welfare, "We Americans must replace our prevailing ethic of expensive self-indulgence with an ethic of rigorous personal responsibility. Government cannot enforce it. Doctors cannot administer it like a drug. We face a choice between taking increased responsibility for our own health or continuing in the present wasteful sick-care system, with its staggering toll in dollars, wasted lives, and grief."

Richard Schweiker, former Secretary of the Department of Health and Human Services, recently insisted that the contemporary scene needs an "urgent health breakthrough . . . that can only take place when we elevate 'wellness care' to the same high level of interest and concern that we've devoted to 'sickness care.' "

We would like to leave you with three thoughts:

1. Toxic substances such as food additives, synthetic chemicals, and polluted air are not going to disappear overnight.

2. The problem cannot be tossed back to the government to be resolved by law, nor can it be left to anyone's conscience; we must continue to educate the public as to what these substances are doing to our bodies and our world.

3. Each person must realize that the diseases to which he is subject are in great part determined by his own actions.

Our freedom of choice, and our right to live as we wish and eat, drink, and breathe whatever we choose, carry a duty and an obligation. We owe it to ourselves and to our society to live in a manner that will promote optimum health.

Our greatest hope for the future is that books such as this one will become unnecessary and obsolete. That will only happen if you strive in tangible ways to make the earth a cleaner planet and to maintain the balance and purity of your body.

Glossary

Addiction: The state of devoting oneself habitually or compulsively to an activity or habit-forming substance, resulting in the inability to stop without experiencing pain or discomfort.

Additive: Any substance added in small amounts to something else, usually food, to produce a desired effect, strengthen, or otherwise alter it.

Adrenalin: See Epinephrine

Aflatoxins: Powerful toxins and carcinogens produced by molds and commonly found in growing crops, foods improperly harvested or stored, and in marijuana cigarettes. Crops commonly contaminated by aflatoxins are grains, legumes, nuts, and seeds.

Allergen: A substance that causes an allergic reaction.

Allergy: An abnormal response to a substance well-tolerated by most people.

Anaphylactic shock: A severe allergic reaction, characterized by loss of consciousness; often accompanied by vomiting, cramps, and swollen throat passages, that makes breathing difficult; suffocation and death may follow.

Antioxidant: A substance that slows down the processes of oxidation; used for detoxification, it offers some protection against air pollution.

Aspartame: A new sweetener, commercially called Nutra-Sweet; laboratory research links it with headaches, depression, and fatigue, and hyperkinesia in children.

Autoimmune disease: An illness sometimes triggered by toxins, which causes the body to react against its own tissue.

Bile: The bitter brown or greenish fluid secreted by the liver that helps the digestion of fats and carries off toxins.

Bioaccumulation: The buildup of chemicals in human and animal tissues.

Biodegradable: Capable of being readily decomposed by biological processes.

Candida: A genus of yeastlike fungi normally found in the body but which can multiply and cause infections, allergic sensitivity, or toxicity.

Carcinogen: A cancer-causing substance.

Challenge-testing: A method of determining sensitivity by touching, eating, or inhaling a suspected substance, and watching for a reaction.

Chemotherapy: The use of drugs or medication in the treatment or control of disease.

Chloracne: Severe acne with pus-filled cysts, often associated with dioxin and PCBs.

Cilia: Tiny hairlike cells that live in the mucous membranes of the respiratory tract and help filter toxins.

Clinical ecology: A new branch of medicine that treats allergic and chemical sensitivities through diet and environmental control, and preferably without the use of drugs. It is also referred to as environmental medicine.

Colonic irrigation: A deep enema purported to aid the body in expelling toxic materials.

Contact dermatitis: A skin rash resulting from touching a substance.

Contactant: A substance that touches the skin.

Detoxification: The process of removing toxic substances or poisons from the body.

Digestive system: A group of organs, including the mouth, stomach, liver, pancreas, and intestines, that changes food and other substances into forms the body can absorb or excrete.

Edema: A swelling of body tissues due to an accumulation of fluids.

Endorphins: A morphine-like chemical produced by the brain and released after strenuous exercise; reported to relieve pain and induce a sense of well-being.

Epinephrine: A powerful hormone released under stress that

stimulates the heart and increases muscle strength and endurance; trade name Adrenalin.

FDA: The U.S. Food and Drug Administration, a government regulatory agency.

Flashback: A sudden reliving of a past sensation or experience, often occurring during detoxification.

Food addiction: An abnormal food dependency with characteristics similar to drug addiction.

Gastrointestinal: Relating to both stomach and intestines.

Glucose: A simple sugar that is easily absorbed into body metabolism.

Herbalism: The practice of treating illness with herbs.

Holistic: An approach to medicine that treats the person as a whole and focuses on prevention, nutrition, living habits, and a positive emotional outlook.

Homeopathy: A branch of medicine based on the theory that a substance that produces disease in a healthy person will also, in minute doses, cure the disease.

Homeostasis: Internal balance; a tendency to stability in the normal physiological state of the organism; good health.

Hyperkinesia: Abnormally increased muscle movement.

Immune system: The mechanism by which the body recognizes a material as foreign to itself and neutralizes, metabolizes, or eliminates it.

Ingestant: Anything swallowed or ingested.

Inhalant: Any airborne substance tiny enough to be inhaled into the lungs.

Irritant: An agent that produces local irritation.

Kidneys: A pair of excretory organs which separate water and waste products from the blood and excrete them through the bladder as urine.

Liver: The body's main organ of detoxification; metabolizes toxins, secretes bile, and performs more than 400 other functions.

Metabolite: The product of a substance broken down by the liver or other body processes.

Metabolization: The transforming of foreign substances into forms the body can absorb or excrete.

Migraine: A severe vascular headache, often accompanied by nausea and vomiting and preceded by visual disturbances.

MSDS: Material Safety Data Sheets; documents that supply information on chemical substances in the workplace.

Mucous membrane: Moist tissue that forms the lining of body cavities which have an external opening, such as the respiratory and digestive tracts.

Naturopathy: A system of treating disease with nutrition, air, exercise, and sunshine, and rejecting the use of medication.

Norepinephrine: A natural body hormone related to epinephrine, and which assists the brain in transmitting nerve impulses.

Organic foods: Foods grown without synthetic chemical fertilizers or pesticides.

Orthomolecular: Pertaining to therapy that advocates good nutrition and the use of vitamins, minerals, and other natural body substances to restore health.

OSHA: Occupational Safety and Health Administration.

Outgas: To give off minute portions of chemical fumes from a solid substance.

Oxidant: A substance that promotes combination with oxygen to form new compounds.

Petrochemical: A synthetic chemical derived from petroleum or natural gas.

Pharmacokinetics: The study of how the body absorbs, metabolizes, distributes, and excretes drugs.

Placebo: An inactive substance containing no drugs, used as a control in an experiment testing the efficacy of a drug.

Pollutant: Any gaseous, chemical, or organic substance that contaminates water, the atmosphere, or the physical body.

Radiation: The combined processes of emission, transmission, and absorption of waves or particles of energy.

Radiation sickness: Illness produced by too much exposure to radiation, as from X rays or atomic explosions, and characterized by nausea, diarrhea, bleeding, loss of hair, and increased susceptibility to infection.

Respiratory system: The system of organs—nose, throat, larynx, trachea, bronchial tubes, lungs—involved in the exchange of carbon dioxide and oxygen between an organism and its environment.

Rotation diet: A diet in which a particular food is eaten only once every four days.

Sebaceous gland: A small skin gland that secretes sebum.

Sebum: The oily skin lubricant secreted by the sebaceous glands.

Solvent: A substance, usually liquid, that can dissolve another substance.

Sweat gland: Any of the many small, coiled tubular glands below the surface of the skin that secrete sweat.

Synthetic: Man-made; not produced normally in nature.

Teratogen: A substance harmful to the fetus.

Terpenes: Natural plant chemicals that can cause severe allergic reactions in the sensitive.

Tinnitus: A ringing in the ears, often a sign of aspirin toxicity.

Tolerance threshold: The maximum amount of toxic substances a person can endure without reacting.

Toxemia: A condition resulting from the distribution of poisonous substances throughout the body.

Toxicology: The science of poisons, their effects, and their antidotes.

Toxic Shock Syndrome (TSS): A severe illness that mainly attacks women using tampons during menses and is characterized by sudden fever, vomiting, diarrhea, and peeling of the palms and soles.

Toxin: Any poisonous or irritating substance.

Tryptophan: An amino acid found in foods and claimed to have a relaxing effect on the brain.

Urinary system: The kidneys, ureter, urinary bladder, and urethra; organs involved in the filtration of toxins and the secretion and discharge of urine.

Withdrawal: Termination of the administration of a habit-forming substance, and the physiological readjustment that takes place upon discontinuation.

Appendix A
General Carcinogenic Substances*

All the cancer-causing substances listed below are commonly found in the environment, although some have been banned and others remain under study.

Arsenic: Compounds are used in pesticides, glass, ceramics, paints, dyes, wood preservatives, and leather processing. More than half a million workers are potentially exposed in metal smelting and in the manufacture and application of pesticides. Consumer exposure occurs in foods, contaminated drinking water, and various products, especially those impregnated with wood preservatives.

Asbestos: It is found in more than 5,000 products, including roofing and flooring materials, brake linings, plastics, textiles, and paper products, and has been used in the fireproofing of thousands of schools and public buildings. The entire population has been threatened to some degree, especially students exposed to deteriorating school construction materials. Asbestos substitutes are being found, but factory workers are still in danger.

Benzene: A major material in the synthesis of chemical compounds, drugs, pesticides, inks, paints, and in the rubber industry, benzene is also used as an octane booster in gasoline. It is found in high concentration in urban areas with heavy truck and auto traffic, and in fruits, fish, vegetables, nuts, dairy products, beverages and eggs.

*Adapted from the *New York Times*, March 20, 1984.

DDT (Dichlorodiphenyltrichloroethane): The whole world has been exposed to this "miracle pesticide," banned in 1974. Traces persist in the food chain, in fatty body tissue, and in the environment.

DES (Diethylstilbestrol): This synthetic sex hormone entered the food chain as a livestock and poultry growth stimulant, and was also used to deter spontaneous abortion in pregnant women. DES in meat and poultry affected the entire U.S. population until its use for that purpose was banned in 1979. The EPA estimates that DES is still prescribed annually for half a million people, mostly men being treated for prostate cancer.

Dioxin: A contaminant in herbicides and defoliants such as Agent Orange, dioxin has been used commercially as a weed killer and in Vietnam. Thousands of plant workers and residents have been exposed through spills. Dioxin remains unregulated while debate continues over the extent of its danger to humans.

EDB (Ethylene dibromide): A chemical additive in leaded gasoline and pesticides, EDB is used as a soil fumigant mainly for citrus fruits. In 1977, NIOSH estimated that 650,000 gasoline station workers were exposed, and most of the population has ingested some EDB in food products. All uses were banned in September, 1984.

Formaldehyde: This gas is colorless, pungent, and found in home building materials, automotive parts, plastics, textiles, embalming fluid, room deodorants, fungicides, cosmetics, and thousands of everyday products. Plant workers continue to be exposed, and the ban on urea formaldehyde home insulation, overturned in 1983, is now on appeal in the courts.

Hair dyes: Beauticians are exposed as well as 33 million women and unknown numbers of men who use home preparations. The FDA proposed regulation in 1979, but the rule was challenged in court and is not yet in effect. Some manufacturers have voluntarily withheld the compound 4-methoxy-m-phenylenediamine (4-MMPD), linked to breast and bladder cancer.

Kepone: This pesticide, carcinogenic in rats, was banned in 1977, but residues still exist in the James River in Virginia, as a result of illegal dumping from a nearby chemical factory.

Nickel: NIOSH estimates that one million U.S. workers are exposed to organic nickel in shipbuilding, aerospace, and a variety of industries where electroplating, ceramics, dyes, paints, magnets, and welding are involved. The EPA claims the entire population may be exposed to low levels in air, food and water. Hazardous waste disposal standards were set in 1980, but industrial regulations are still under study.

Nitrites: Mainly used as a food and antibacterial preservative, nitrites are common in smoked meats such as bacon, hot dogs, and cold cuts. In the human digestive tract, they combine with other substances to form nitrosamines, which are carcinogenic. Exposure can be reduced by limiting intake of cured meats.

PCB's (polychlorinated biphenyls): These synthetic compounds were mainly used in electric transformer coolants, hydraulic fluids, coatings, sealants, and pesticide extenders until banned in 1979. They still persist in the environment, in fish, cheese, eggs and meat, in water, soil, air, and human tissue.

Saccharin: When cyclamates were banned as a carcinogen in 1969, saccharin took over as an artificial, nonnutritive sweetener. Linked with cancer and leukemia in rats, it must now carry a warning label, but is still apprcved by the FDA and still being studied.

Soots, coke, and tars: Coke, or carbonized coal, is used to extract metals from their ores. OSHA estimates more than 131,000 workers are exposed to coal-based tars and pitches that have shown "overwhelming" evidence of carcinogenicity. Environmental contamination is widespread and affects us all. The EPA is considering emission controls in factories.

Tobacco: Cigarette smoke has been directly associated with lung cancer, and the latest government estimate indicates cigarette smoking as the cause of one-third of all cancer deaths. Cigarettes must carry a warning label, and advertising is banned on radio and television, but the Surgeon General reports that one-third of the population still smokes.

TRIS [(hydroxymethyl)-aminomethane]: A flame retardant in children's sleepwear, toys, and furniture, Tris was banned as carcinogenic in 1977 and has not been produced since then. Millions of children, however, were exposed by skin absorption or orally by toys and garments placed in the mouth.

Vinyl chloride: The raw material of plastics and formerly a propellant in aerosol containers, vinyl chloride has been closely linked with a rare liver cancer in factory workers. OSHA imposed workplace regulations in 1974, but the FDA is considering withdrawing a 1975 order regulating vinyl chloride in food packaging, which manufacturers claim is now safer.

Appendix B
Medication and Its Effects on Pregnancy

All drugs are best avoided in pregnancy unless absolutely necessary. On the following pages, generic names appear before trade names. Those listed are FDA approved.

Other factors affecting the action of a drug or its metabolites on the fetus include gestational age at the time of drug exposure, drug dosage, and route of administration, and the genetic predisposition of the fetus to respond to a particular drug.

QUESTIONABLE (These drugs may be safe, but have not been adequately tested on the human fetus. Use with caution.)

Antacids for stomach upsets: aluminum hydroxide and magnesium hydroxide (Di-Gel, Gelusil, Maalox, Mylanta, Simeco), calcium carbonate (Alka-2, Dicarbosil, Mallamint, Titralac, Tums), sodium bicarbonate or baking soda (Alka-Seltzer Gold, Bisodol powder). Do not take any antacid for more than two weeks. Preferable to use products in the first group (Di-Gel, etc.) rather than antacids high in sodium.

Anticoagulants: heparin (Calciparine, Hepalean). Heparin does not cross the placental barrier, but its teratogenic potential is still unknown. Use with particular caution during the last trimester.

Anticonvulsants: primidone (Mysoline); possibly associated with congenital malformation in the offspring of epileptics, but no definite cause and effect relationship has been proven.

Antidiarrheals: diphenoxylate and atropine (Lomotil), pare-
goric products; may lead to drug dependency in
newborns.

Antiemetics for vomiting: meclizine (Antivert, Bonine, Elde-
zine) cyclizine (Marezine), trimethobenzamide (Tigan).
High doses of meclizine have been linked with cleft
palates in rats, but studies have found no correlation
with humans. Rabbits receiving trimethobenzamide
showed some loss of embryonic tissue and an increase
in stillborns. Relevance to humans is unknown.

Antihistamines: most can be passed to fetus and to breast-fed
baby. Some decrease mother's milk production.

Anti-infection drugs: penicillin has been safely used for two
decades, but no studies have measured the fetal effects
of newer synthetic penicillins (methicillin, dicloxacillin,
and ampicillin). All penicillin drugs pass through the
mother's milk to the infant. Persons allergic to penicillin,
of course, should never take any form of the drug. Other
anti-infectives, cephalosporins (Ceclor), nystatin (Can-
dex, Mycostatin, Nilstat) appear to be well tolerated
during the second and third trimesters of pregnancy, but
the question of teratogenicity remains unanswered.

Anti-inflammatory Agents: fenoprofen (Nalfon), ibuprofen
(Motrin), oxyphenbutazone (Oxalid, Tandearil), phen-
ylbutazone (Azolid, Butazolidin); tolmetin (Tolectin);
can cause salt retention and bloating. Studies in animals
show possible embryotoxicity from oxyphenbutazone.

Antipsychotics: droperidol (Inapsine), haloperidol (Haldol),
thiothixene (Navane). In a French study, neonates exposed to these drugs had a higher rate of limb malformations than did the controls; an American study
demonstrated no increase in congenital malformations.
In view of contradictory findings, caution is urged.

Cardiovascular or blood vessel drugs: methyldopa (Aldomet),
hydralazine (Apresoline, Dralzine), chlorothiazide
(Chlorulan, Diuril), digitalis (Digoxin, Crystodigin, Lan-

oxin), propranolol (Inderal). No obvious teratogenic effects have been reported but methyldopa does cross the placenta and appear in umbilical cord blood. Hydralazine has caused facial malformations in animals, but no adverse effects in humans have been established. Propranolol has been linked with neonatal respiratory depression, hypoglycemia and, rarely, with intrauterine growth retardation. Propranolol dosages under 160 milligrams a day, however, have little observable effect on the fetus.

Painkillers: aspirin; use sparingly, and avoid in last three months as it can complicate delivery. Acetaminophen, codeine (Alorain, Codalan, Tega-Code-M), ethoheptazine (Equagesic); meperidine (Demerol, Mepergan); oxycodone (Percodan); pentazocine (Talwin); propoxyphene (Darvon) and all painkillers taken in the third trimester may prolong pregnancy or labor. Read labels carefully to see what other chemicals product contains (discard any painkiller with phenacetin; it may cause kidney damage and anemia).

Sedatives or sleep medicine: barbiturates (Nembutal, Seconal), chloral hydrate (Aquachloral, Noctec), ethchlorvynol (Placidyl), flurazepam (Dalmane), glutethimide (Doriden); methaqualone (Meguin, Quaalude), methyprylon (Noludar), triclofos (Triclos). The lack of adverse effects on fetal development has not been established, therefore their use in pregnancy is not recommended. Some may cause dependency and withdrawal symptoms in the newborn.

UNSAFE (There is adequate evidence to indicate that these drugs may damage the fetus.)

Anticancer: aminopterin: busulfan (Myleran), azathioprine (Imuran), chlorambucil (Leukeran), cyclophosphamide (Cytoxan), fluorouracil (Fluorouracil, Adrucil), mercaptopurine (Purinethol), methotrexate (Methotrexate, Mexate), and thioguanine (Thioguanine) are extremely potent drugs and should be avoided during pregnancy, particularly in the first trimester, unless there is ample evidence that the benefits can be expected to outweigh the risks. Any drug designed to kill malignant cells may also kill healthy cells in the embryo.

Anticoagulants: warfarin (Coumadin, Panwarfin). Use of warfarin during the first trimester appears to be associated with a variety of facial and optical defects, mental retardation, and other central nervous system abnormalities.

Anticonvulsants: diphenylhydantoin (Dilantin), carbamazepine (Tegretol), ethosuximide (Zarontin), methsuximide (Celontin), trimethadione (Tridione), valproic acid (Depakene). Dilantin has been found in breast milk in amounts large enough to produce undesirable effects in nursing infants. Higher incidences of birth defects have been reported in mothers taking antiepileptic drugs, but data are insufficient.

Anti-infection drugs: erythromycins (E-Mycin, Ilotycin, Wyamycine) are associated with liver dysfunction; tetracyclines (Declomycin, Vibramycin, Vibra-Tabs, Achromycin) may cause tooth discoloration in newborns; trimethoprim (Proloprim, Trimpex), and metronidazole (Flagyl) may cause bacteria mutations in the first trimester and should be avoided at that time. Nitrofurantoin (Furadantin) can cause anemia in newborns.

Anti-inflammatory Agents: indomethacin (Indocin), naproxen (Anaprox, Naprosyn) should be avoided in the third trimester because of possible effects on the fetal cardiovascular system.

Antipsychotics: lithium carbonate (Eskalith, Lithane, Lithonate), trifluoperazine (Stelazine), chlorpromazine (Thorazine), perphenazine (Trilafon), prochlorperazine (Compazine), thioridazine (Mellaril). The use of lithium in the first trimester increases the risk of a rare cardiac lesion, Ebstein's tricuspid valve anomaly. The incidence of lithium-associated congenital anomalies, is, however, under ten percent. Chlorpromazine can cause jaundice and muscle tremors in newborn.

Hormones: diethylstilbestrol (DES) in first trimester can cause genital anomalies and cancer in both male and female fetus. Several reports link estrogens and progesterones in pills, creams, and suppositories with fetal heart defects and limb reduction defects. Stop all hormones and birth control pills at first suspicion of pregnancy.

Tranquilizers for nerves: meprobamate (Equanil, Miltown), diazepam (Valium), and chlordiazepoxide (Librium) have been associated with congenital malformations during the first trimester of pregnancy, and should be avoided. One study reported in *Drug Therapy*, April 1983, indicated that women who smoke and use tranquilizers (Valium or Librium) run a 3.7 times higher risk of delivering a malformed infant than women who smoke but do not use tranquilizers.

Used late in pregnancy, tranquilizers may cause breathing problems in newborns.

Appendix C
Probable Teratogens

Teratogens are substances harmful to the fetus.

Metals

Arsenic. This element possesses both metal and nonmetal properties. It is found in such products as fungicides, pesticides, dyes, paints, veterinary drugs, and cigarette smoke. Experiments have demonstrated birth defects in animals.

Cadmium. Traces are found in polluted air, especially in highway dust with its high concentration of particles from auto tires and batteries. Cadmium is often leached from solders and pipes into hot water and is known to cause birth defects in rodents.

Gold. Gold-containing compounds (Myochrisine, Solganal), prescribed for juvenile-onset rheumatoid arthritis, should be discontinued three to six months prior to pregnancy.

Lead. Lead crosses the placenta and can be responsible for congenital retardation and stillbirths. In one study, mild brain damage in newborns was attributed to lead in illegal whiskey that the mothers drank during pregnancy. It is mainly found in gasoline, batteries, pottery glazes, and house paints.

Mercury. It was first noted in Japan, in 1953, that pregnant women who ate fish contaminated with methyl mercury gave birth to brain-damaged babies. Today, methyl mercury is a proven teratogen.

Chemicals

Acrylonitrile. Used in the production of plastics, this liquid is found in food-packaging materials, pesticides, synthetic fibers and plastics, foods, and fish. High doses cause deaths of embryos in animals.

Aldicarb. An active ingredient in the insecticide Temik, aldicarb is a nerve poison ten to fifteen times more potent than cyanide. In 1979, contaminated communities in Long Island, New York, showed abnormally high miscarriage rates.

Benzene. Chromosome damage and birth defects in rats have been attributed to benzene. It is found in such consumer products as paint solvents, glues, cleaning fluids, hobby products, and cigarette smoke.

Chlordane. Residues of this pesticide used in termite control accumulate in body tissue and are frequently found in umbilical cord blood and mother's milk. It is reported to cause leukemia in infants born to mothers who were heavily exposed during pregnancy.

Dioxin (TCDD). A contaminant in herbicides and Agent Orange, the notorious defoliant used in Vietnam, this is the most toxic known chemical and is still found in many commercial weed killers. In tiny concentration, dioxin has been shown to induce birth defects such as cleft palate and malformed limbs.

Ethylene dibromide (EDB). This well-known pesticide, recently found in many grain products, and in citrus fruits sprayed with the chemical, has been linked to birth defects in animals.

Polychlorinated biphenyls (PCBs). These compounds accumulate in body tissue and specifically in mothers' milk. Pregnant women exposed to PCBs, which are used to protect and insulate electrical equipment, have experienced spontaneous abortions, stillbirths, and undersized infants.

BHT, EDTA, PVP. These and other ingredients in foods and cosmetics are strongly suspected to be teratogens. (See Appendix D.)

DBCP. In 1977, it was discovered that men who had worked with this pesticide were sterile, leading scientists to conclude that DBCP is a reproductive hazard to both sexes.

Toluene. Women who used this chemical in varnishing work in factories reported premature births and newborns who acted drugged. It is commonly found in spot removers and cigarette smoke.

Vinyl chloride. In several cities in Ohio, factory exposure caused central nervous system malformations in newborns. It is found in vinyl products, cigarette smoke, and in foods wrapped in heat-sealed plastic.

Radiation

The fetus is particularly vulnerable to radioactive materials. During pregnancy, X rays and even low-level radiation exposure should be avoided. Damage to the infant in utero could result in childhood cancer, leukemia, mental retardation, and small head size.

Personal Habits

Additives. Select foods, cosmetics, and other contactants with as few additives as possible. A common additive such as monosodium glutamate (MSG) has been shown to cause reproductive dysfunctions in laboratory animals.

Alcohol. There is no safe level of alcohol during pregnancy. Recent ultrasound studies show that a single cocktail can severely disrupt fetal chest wall movements, or "breathing." Every time the mother drinks, the baby takes a drink, too. Alcohol reaches the same concentration in fetal blood as in the mother's.

The 1982 Merck Manual warns, "Maternal alcohol abuse throughout pregnancy is felt to be the most important single cause of drug-induced teratogenesis. The most serious consequence of Fetal Alcohol Syndrome (FAS) is severe mental retardation."

Coffee. Most doctors urge a limited intake, since the alkaloid (caffeine) crosses the placenta to the fetus and is known to cause birth malformations in animals. The FDA advises pregnant women to avoid all products containing caffeine.

Nicotine. Smoking is associated with miscarriage, stillbirth, reduced birth weight, which increases risk of mental and physical abnormalities, as well as fetal and infant mortality. Even cutting down on cigarettes would benefit the fetus of the woman who feels she cannot quit completely.

Recreational drugs. Pregnant women dependent on any illicit drug, from marijuana to heroin, are subject to premature deliveries; their offspring are subject to low birth weights and infant morbidity. These women generally require more anesthesia during labor and may give birth to addicted babies who have to go through severe withdrawal. Many infants do not survive.

Toxoplasmosis. This disease is caused by a fairly common parasite and can damage the fetal brain and nervous system if acquired during the first trimester. The parasite appears in cat (and only cat) stool and in raw or undercooked meat. Many adults are immune to toxoplasmosis because at one time or another, they had a case too mild to detect; nevertheless, pregnant women should not handle cats, work in gardens where cat excrement is present, or empty cat litter boxes.

Vitamins

Taking vitamins during pregnancy is either essential, useful, harmless, or teratogenic, depending on your source of information. Many obstetricians prescribe a safeguard supplement to cover the minimum daily requirements. In normal pregnancies, this should be sufficient.

Taking megadoses may expose both the mothers and their unborn babies to bodily harm. The information that follows is the latest.

Vitamin A. This is toxic at 100,000 International Units (IUs) and can cause headaches and lethargy.

Vitamin B₆. Doses of 2,000 milligrams daily over a period of four months can damage the nervous system, resulting in numb limbs and loss of muscle coordination. High doses of pyridoxine have been associated with withdrawal seizures in the baby.

Vitamin C. Ascorbic acid may cause "rebound" scurvy in a newborn. It happens when an infant has become accustomed to megadoses—three to ten grams daily—of vitamin C which suddenly stop during labor. Dr. Linus Pauling, however, recommends at least one gram a day to "alleviate morning sickness and build up fetal immunities. Vitamin C is a natural substance that also benefits women at risk of miscarriage." Always drink large amounts of fluids with 3 or more grams of vitamin C a day.

Vitamin D. Pregnant women should be especially careful of doses above 10,000 to 25,000 IUs. Too much vitamin D can lead to calcification of the kidneys and blood vessels.

Vitamin and mineral supplements. Some substances in these preparations, such as iodides, may adversely affect the thyroid of the fetus, especially if taken during the second or third trimester.

Vitamin K. In the last trimester, large amounts of vitamin K may increase the severity of jaundice in infants born with the disease.

Appendix D

Toxins in Cosmetics—
To Be Used
Minimally
or Avoided

Replace all preparations that have a warning on the label.

Acetone. Found in nail polish remover; affects nervous system.

Blue dye No. 1. Found in aftershaves and male colognes, body creams, soaps, toothpastes, blushes, lipsticks; suspected carcinogen.

Butylated hydroxytoluene (BHT). A phenol derivative corrosive to skin, found in lipstick, baby oils, eyeliner pencils, soaps; suspected teratogen.

Coal tar. Found in skin-care products and medicated shampoos; may cause skin allergy, cancer.

Calcium disodium edetate (EDTA). Found in eye shadows, soaps, skin softeners; suspected teratogen.

Estrogen derived from human placenta. A hormone found in face and body creams and hair lotions; may cause changes in female reproductive organs; may cause males to develop female characteristics; potential carcinogen.

Formaldehyde. Common in deodorants, antiseptic shampoos, hairsetting lotions, mouthwashes, nail polish, perfumes, and toothpaste; causes skin sensitivity, respiratory tract and eye irritation; a carcinogen in animals.

Green dye No. 6. Found in pine shampoos, mint-flavored preparations; usually contains a small amount of p-toluidine, a carcinogen.

Iron oxides. Found in eye makeups, rouges; suspected carcinogen.

Lead acetate. Found in hair dyes; is absorbed into skin, causes cancer in animals.

Polyvinylpyrrolidone (PVP). Found in eyeliners and setting gels; suspected carcinogen.

Red dye No. 2 and Red dye No. 4. Banned by FDA as carcinogenic; a good reason not to use cosmetics over a year old, particularly pink or red ones.

Toluene. Found in nail polishes; attacks nervous system and may cause hearing loss; suspected carcinogen.

Triethanolamine (TEA). Found in moisturizers, perfumes, body and suntan lotions, bubble baths, hair gels, shaving foams; may combine to form N-nitrosodiethanolamine (NDELA), a potent animal carcinogen.

Appendix E

Home and Office Products Toxic to Skin

Toxic damage is usually temporary and heals when use is discontinued.

Abrasive powders. Cleaning detergents contain phenols, formaldehyde, benzene, and ammonia, and can cause irritation, reddening, blistering, or skin destruction.

Air fresheners. Carbolic acid and formaldehyde, found in some fresheners, can burn and destroy tissue when they touch the skin.

Antiseptics. Thimerosol and merthiolate are common mercury salts that act as skin irritants.

Aquarium products. Metal salts such as aluminum, copper, nickel, tin, and zinc can cause lesions or burns.

Bath soaps. Many contain nitrosamines, an animal carcinogen, which can be absorbed through the skin.

Batteries. Even dry cells can explode or leak acid, causing skin burns.

Colored dyes. These contain aniline, a colorless, irritating oil also found in resins, varnishes, and stamp pad ink. Aniline reacts with the blood and may turn the fingers, lips, and whites of the eyes a bluish color.

Crayons. Offenders include dyes, waxes, lead chromate; repeated skin contact can lead to eczema and slow-healing ulceration.

Detergents. The complex chemicals in laundry formulas sometimes cause dermatitis, or severe optical damage if there is eye contact.

Dishwashing liquids. These compounds contain glycerin and calcium sodium edetate (EDTA), are highly soluble, and easily penetrate the skin, where they can cause various degrees of damage. Many contain phosphate, also a skin irritant.

Disinfectants. Cresol, an ingredient in germ-killers, can be absorbed through skin and attack the liver, kidneys, spleen, pancreas, and central nervous system.

Drain cleaners. Common ingredients are sulfuric acid and lye, which can burn and discolor skin, with slow healing and scars.

Electrical transformers. Many are insulated with PCBs which, when handled by workers, can cause skin rashes, as well as systemic symptoms such as headaches, swollen joints, nausea, and vomiting.

Fireworks. The main toxins are arsenic, mercury, antimony, and phosphorous; all can produce skin pustules and burns as well as severe internal problems.

Ink eradicators. The active ingredient is sodium hypochlorite, a strongly irritating bleach.

Lead. This common metal is found in batteries, beads, fake pearls, gasoline, some glass, magazine and newspaper ink, paint, pottery, rubber, solder, toys, and many more products. Skin absorption induces apathy, fatigue, loss of appetite, and eventual brain cell damage.

Leather. Contact with the chromium used in the tanning process can lead to incapacitating dermatitis.

Liquid wall or furniture cleaners. Some contain sodium phosphates which penetrate the skin and cause rashes.

Moth repellant. Continued handling of these products, which contain naphthalene, may produce an itchy, weepy, crusty dermatitis. Eye contact injures the cornea.

Oven cleaners. The main cleaning ingredient is lye, which can burn the skin and cause internal poisoning.

Paint, lacquer, and varnish remover. Benzene, methyl ethyl ketone, and other compounds cause skin scaling and cracking.

Paper products. Using facial tissue and paper towels that contain formaldehyde (for strength and water-resistance) may trigger severe rashes. Be alert to the possibility and change brands if you break out; label information, unfortunately, doesn't reveal ingredients.

Photographic supplies. These can include such chemicals as turpentine, ammonium hydroxide, phenol, methanol, chloride, formaldehyde, hydrochloric and sulfuric acid. They all irritate skin and eyes and can provoke a red rash or acid burns. Methylene chloride and benzene may cause cancer.

Polyester, polyurethane, styrene, etc. Almost all synthetic materials are skin irritants and mild toxins; most people can metabolize small amounts with no reaction, but if contact continues, volatiles from the fibers can be absorbed and cause central nervous system disorders.

Rubber cement. Toluene is the harmful component; it passes through the skin and may induce dizziness and intoxication.

Shoe-cleaning products. Many contain nitrobenzene. Easily absorbed into the skin, it can cause blue discoloration, shallow breathing and vomiting.

Spot removers. These are made of many toxic chemicals, including toluene, benzene, and sulfamic acid, which may irritate the skin and cause central nervous system symptoms.

Thermometers. If a thermometer breaks, liquid mercury can be absorbed through the skin and cause headaches, pain in the extremities, and mood and behavioral changes.

Turpentine. A common ingredient in furniture polish, this is a volatile oil of ethers, alcohols, and hydrocarbons that evaporates at room temperature and is easily absorbed by the body.

Watercolor paints. These contain gum resins which may cause cell damage when applied to the skin.

Appendix F

Intentional Food Additives to Avoid

Artificial colorings. Blue No. 1 and Blue No. 2, used in candy, pastry, and beverages; causes cancer in rats.

Citrus Red No. 2, used to color oranges, but does not seep through to the pulp; causes cancer in mice.

Green No. 3, used in candy and beverages; is poorly tested and may be carcinogenic.

Orange B, used in hot dogs; causes cancer in mice.

Red No. 3, used in canned cherries, candy, gum, and baked goods; induces hyperactivity in children, and may cause cancer.

Red No. 40, used in soft drinks, sausage, pet food, and junk foods; may cause cancer in mice.

Yellow No. 5, tartrazine, used in gelatin desserts, soft drinks, and candy; is a potent allergen and possible carcinogen.

Benzoic acid. Used as flavor preserver and antimold agent; acts as a mild irritant to skin, eyes, and mucous membranes; toxicity increases when combined with other additives.

Brominated vegetable oil (BVO). Gives cloudy appearance to citrus-flavored soft drinks; toxic residues of BVO store in body fat.

Butylated hydroxyanisole (BHA). Retards rancidity in ice cream, oils, meats, and processed meats; a 1983 report by Oregon State University scientists claims that it protects the human liver from toxic chemicals; known to cause allergy, affect liver and kidney function, increase incidence of stomach cancer.

Butylated hydroxytoluene (BHT). Antioxidant in some cereals and potato chips; credited by some scientists for declining incidence of stomach cancer; same negative effects as BHA but more toxic to liver; Colorado researcher Dr. Alvin M. Malkinson's 1983 studies show that BHT promotes lung tumors in mice; accumulates in human fat.

Calcium (or sodium) propionate. Mold inhibitor in breads and rolls; provokes reactions in allergic persons; inhibits assimilation of calcium.

Carrageenan (seaweed). Used as gel; causes ulcers, liver lesions, and birth defects in animals.

Cellulose gum (sodium carboxymethylcellulose). Thickener and lump preventer; causes cancer in mice.

Calcium disodium edetate (EDTA). Widely used to retain food color and flavor; in large amounts, causes kidney damage and gastrointestinal problems.

Flavorings. Approximately 1,500 artificial flavors exist; each one is usually a combination of compounds. The words "added flavoring" or even "almond flavoring" on a label give no indication of the chemicals involved. Tests on animals revealed adverse effects from many flavorings.

Hydrogenated or partially hydrogenated fat. Liquid oil made into hard or semihard fat, then bleached and deodorized; used to resist deterioration; as a saturated (hard) fat, it elevates cholesterol and is strongly linked to heart disease.

Hydrolized vegetable protein. Flavor enhancer; causes nerve damage in infant mice.

Mannitol, Sorbitol, Xylitol. Sweeteners; have laxative effect in large doses.

Modified starches. Used as emulsifiers and food "stretchers"; high levels cause abnormalities in rats.

Monosodium glutamate (MSG). Flavor enhancer; contains wheat, corn, and beet sugar by-products which are common allergens; may be contaminated with aspergillus mold; destroys brain cells in every species of lab animal, especially newborn; can cause headache, chest tightness, burning sensation in neck and arms, and possible seizures.

Propyl gallate. Retards spoilage of fats and oils; poorly tested; increases risk of cancer in mice.

Quinine. In tonic and quinine water, bitter lemon; can cause vomiting, disturbances in vision, ringing in ears; related to birth defects.

Saccharin. Coal tar sweetener in sodas, diet foods, toothpaste; promotes bladder cancer in rats; acknowledged by government as carcinogen, but still on the market because of pressure from food industry.

Silicates. Used to absorb moisture and keep powders and salts free-flowing; some are a form of asbestos with carcinogenic potential.

Sodium bisulfite. Used as preservative and antidiscolorant in fresh fruits and vegetables, shellfish, and wines; destroys vitamin B_1 (thiamine); produces gastric distress and diarrhea; is a potent allergen capable of causing anaphylactic shock, the most severe allergic reaction, characterized by a drop in blood pressure, vomiting, and swollen throat passages.

Sodium chloride (salt). Seasoning and preservative; less than one-tenth of a teaspoon a day keeps the body running smoothly. The average American eats more than twenty times that amount, increasing susceptibility to edema, kidney problems, hypertension, and high blood pressure. Sea salt does not contain the dextrose, potassium iodide, or aluminates present in table salt but may be contaminated by seagull droppings or oil slicks.

Sodium nitrate. Same uses as sodium nitrite (see below); readily converts to sodium nitrite in the body.

Sodium nitrite. Preservative and color fixative for cured meat and fish; combines with other substances in the digestive tract to form nitrosamines, strong carcinogens, which can also trigger diabetes attacks.

Sulfur dioxide. Antioxidant and bacterial inhibitor for wines and dried fruits; causes gastric distress, headaches, and diarrhea; destroys vitamins A and B_1.

Appendix G
Kitchen Utensils: Toxic and Otherwise

Aluminum ware. Aluminum salts are toxic to brain cells and have recently been linked to Alzheimer's disease; whether enough seep into foods to warrant concern is not known. Individuals sensitive to aluminum should avoid cooking with it, using foil wrap, or storing food—particularly acidic fruits such as pineapple—in aluminum containers.

Cast-ironware. Iron interacts with food the same way aluminum does, but iron salts are not toxic; on the contrary, they are a beneficial nutrient and helpful for people with iron-deficiency anemia. Cast-iron cookware is also low-cost, durable, and efficient.

Clay pottery. Glazes used to coat earthenware may contain lead, cadmium, and other toxic metals which can leach into food. The leaching is greater when the pottery interacts with acid-containing products, such as tomato sauces, coffee, wine, and fruit juices; heating also increases metal migration. Good quality commercial crockpots are generally safe; avoid homemade mugs or pottery, and be especially wary of imported clay products which often contain excessive levels of toxic metals.

Copper pots. These are excellent heat conductors, but should be lined with tin or stainless steel; otherwise, copper migrates into food and can cause toxicity.

Decorated glasses. Drinking glasses with decals containing high levels of lead and cadmium can be toxic; children bite, chew, or lick the designs, or touch them with sticky fingers that also touch their mouths and food.

Enamel ware. Buy the best quality. Cheaper dishes and containers made with soft enamel can leach toxic oxides into food.

Glass. Glass containers do not conduct heat well, but they are chemically inert and leach no metals or chemicals; better for food storage than plastic.

Nonstick coatings. Good quality pots and pans with Teflon, SilverStone, or other silicone resin coatings are generally nontoxic, though they release minute quantities of fluoride. Many people use liquid lecithin to coat pans; this is safe and even has nutritional value but is sticky and messy to use.

Plastic products. Chemicals in plastic and styrofoam are volatile and do leach into food. The convenience of cooking in a bag, or using plastic soft drink bottles or plates is not always worth the additives you're getting. Most china, glass, or porcelain containers are preferable to those made of plastic, wood, metal or paper.

Stainless steel. Stainless steel is almost completely inert and nonhazardous, and should cause no problems unless a person is metal-sensitive.

Wood products. These contain no toxic metals but may have vegetable resins—plant substances used in lacquers and varnishes—which outgas. Also, wood is particularly susceptible to bacteria and molds and must be kept clean at all times.

Appendix H
Common Water Contaminants

Arsenic. Pesticides and improperly disposed chemicals in ground water carry arsenic into the water system. In large quantities, arsenic can cause abdominal pain, birth defects, brain damage, liver and kidney disease, skin and lung cancer.

Asbestos. Dangerous dust comes from asbestos-containing rocks, cement pipes, and industrial wastes dumped in waterways and landfills. Particles from brake linings and fragments of auto tires are washed from city streets by rain, pass through sewage plants untouched, and turn up in home supplies. Strong evidence links asbestos with lung and gastrointestinal tract cancer, as well as the rare, fatal form known as mesothelioma.

Benzene. The water near gasoline refineries and many other manufacturing plants contain high levels of this solvent. It causes cancer and leukemia in rats and damages human blood cells.

Carbon tetrachloride. Fortunately, this chemical is no longer widely used as a cleaning solvent, but much of it remains in the environment. It is known to cause liver, kidney, and central nervous system damage and is a suspected carcinogen.

Chlorine. Used extensively to destroy bacteria and to reduce the spread of infectious disease, chlorine also combines with plant acids from organic matter like leaves and molds and produces compounds such as chloroform, a carcinogen. Chlorine is a potent allergen and a toxin in itself.

Dichlorodiphenyltrichloroethane (DDT). Banned in 1972, residues of this long-lasting pesticide are still prevalent in agricultural water supplies; they accumulate in breast milk and body tissue, where they can cause liver damage and possibly cancer.

Dioxin (TCDD). A contaminant in some defoliants such as Agent Orange, used during the Vietnam war, Dioxin has been strongly implicated in a high incidence of birth defects, skin rashes, and liver damage. It is found in water supplies around farm areas sprayed with herbicide, and in fish caught in such lakes and rivers.

Extractables from pipes. Fluoride, sulfur, plastic, copper, lead, zinc, cadmium, and cobalt all leach into drinking water from pipe erosion and cause a variety of physical symptoms. Unfortunately, no pipe has yet been manufactured of completely inert materials. Letting water run before using it cuts down on pipe contamination.

Mercury. Certain microorganisms in water convert this chemical to the highly toxic methyl mercury; it is passed on to humans by fish and can cause brain damage, birth defects, and death.

Nitrates. Residues containing nitrates come from agricultural runoff that flows from croplands directly or indirectly into water supplies. Nitrates form carcinogens called nitrosamines in the human digestive tract.

Polychlorinated biphenyls (PCBs). Although banned, PCBs persist in fresh water and fish. They accumulate in the human body where they can cause birth defects, liver and skin problems, and chronic fatigue.

Toluene. Because this solvent has not proven to be a carcinogen, high levels are allowed in water. It has been shown to cause behavior and emotional disorders, heart, liver, and kidney damage.

Trichloroethylene (TCE). A solvent with numerous applications, TCE is used as metal degreaser, dry-cleaning agent, and for decaffeinating coffee. It seeps into water supplies from industrial usage and can cause central nervous system damage, heart and liver malfunctions, and cancer in mice.

Vinyl chloride. The most widely used material in the production of plastic, this chemical is found in dangerous amounts in water near factory sites. It can cause a specific type of liver cancer, and possibly breast, brain, and lung cancer.

Appendix I
Indoor Air Pollutants

Ammonia. A colorless gas with a strong, familiar odor, this compound is widely used in cleaning products and some office machines. It irritates eyes and respiratory passages; long-term effects are unknown.

Asbestos. This pernicious material is used in fireproofing, electrical insulation, acoustical tile, backing of vinyl floor tile, and in more than 3,000 consumer products. Auto mechanics routinely repair brakes with asbestos parts, construction firms still build with it, and great quantities spew into the air during the demolition of old buildings. Airborne fibers cause asbestosis, a chronic lung ailment, as well as chest, colon, stomach, and other cancers that usually become discernible only when it is too late to cure them. Lung disease is especially prevalent in homes down-wind of asbestos factories. Many medical authorities recommend a complete ban on all uses of this mineral.

Benzene. A volatile component of gasoline and a probable carcinogen found in synthetic fibers, spot removers, plastics, paint solvents, and tobacco smoke, low concentrations of benzene can irritate the liver, kidney, and gastrointestinal tract. Higher doses act on the central nervous system, causing behavior changes and drowsiness, and can damage blood cells and bone marrow.

Combustion products of natural gas and other fuels. These toxic combinations of carbon monoxide, nitrogen oxides, and other chemicals come from gas kitchen stoves, gas or fuel-oil heaters, gas-burning appliances such as water heaters and refrigerators, warm air furnaces, and gas pipes. Car exhaust fumes often rise into homes from

192

basement garages. Traffic fumes can blow through open windows. Symptoms of gas toxicity include respiratory and heart problems, visual distortions, sleepiness, mood changes, and mental disorders. A malfunctioning furnace often leads to overt illness.

Ethanol. Also known as ethyl alcohol, this compound is found in plastics, office fluids, medications, and cleaning products. Inhalation can cause cerebral symptoms such as sleepiness, confusion, and headaches.

Ethylene oxide. Hospitals and medical-supply manufacturers use this gas to sterilize instruments. It can cause skin, eye, and throat irritations, dizziness, headaches, and in large doses, unconsciousness. It is also a suspected carcinogen.

Fluorocarbons. These chemicals are not highly toxic in themselves, but when used in aerosol propellants, fluorocarbons combine with insecticides, perfumes, and so on, remain in the air, then settle into carpets, clothing, and furnishings. Sprays create clouds of chemicals that are inhaled and ingested into the bloodstream, where they cause symptoms of toxicity.

Formaldehyde. Much media attention has focused on this chemical and with good reason. It is commonly used in concrete, stone, treated lumber, as a glue in plywood, in the particle board found in kitchen cabinets and mobile homes, and in other building materials. Urea-formaldehyde foam insulation (UFFI) is a foamy material pumped into the walls of homes and buildings and left to harden. In large amounts or improperly installed, it can cause toxic symptoms of headaches, nausea, behavior changes, respiratory difficulties, rashes, and influenzalike symptoms. Although the pungent gas causes cancer in laboratory animals, its long-term effects on humans are unknown.

This highly toxic chemical is also found in adhesives, disinfectants, deodorants, soap, toothpaste, shampoo, air fresheners, spray starch, dyes, draperies, synthetic carpets, concrete, plastic, paper, cloth, wood, leather, and almost every manufactured product you can name. Levels of formaldehyde in sealed office buildings often exceed standards known to be safe. Although outlawed in Canada as a home insulator, it is still allowed in the U.S.

Lead. One of the oldest known contaminants, lead is found in numerous household products and foods. It accumulates in the body in bone and soft tissue and affects the liver, kidneys, and reproductive and nervous systems. Even low doses can cause brain damage and behavior disorders. It causes cancer in rats.

Methanol. A common irritant, methanol—also known as wood alcohol—escapes from copy machines. It affects the skin, lungs, eyes, and central nervous system, and in large doses, can damage the liver.

Ozone. A pungent "electric" odor identifies this bluish gas given off by office copy machines and used as a disinfectant in water supplies and industrial wastes. (See Outdoor Air Pollutants.)

Polychlorinated biphenyls (PCBs). Widely used as fire retardants and insulators on electrical transformers, PCBs seep into the air when the equipment is damaged especially by fire. Studies indicate they induce liver damage and cancer in rats. The most significant human effect is a pus-filled temporary dermatitis called chloracne; others who have been heavily exposed complain of nausea, swollen joints, headaches, loss of appetite, sore mucous membranes, and jaundice.

Perchloroethylene. Almost all dry cleaners use this suspected carcinogen. The dry-cleaning industry has lobbied forcefully and successfully against regulating its use.

Radon. This radioactive gas is produced by the decay of radium and occurs naturally in soil and rocks, hence is found in brick, cement, granite, concrete, and other building materials. Studies have established a firm link between radon and lung cancer.

Vinyl chloride. Plastic products, lighting fixtures, upholstered furniture, wires in electric blankets, pipes, and synthetic carpets all contain this compound. It is a known carcinogen and strongly suspected to cause gene mutation, birth defects, chronic bronchitis, bone disorders, and ulcers.

Appendix J

Outdoor Air Pollutants

(See also Appendix I: Indoor Air Pollutants and Appendix H: Common Water Contaminants)

Carbon monoxide. About 80 percent of this colorless, odorless gas comes from the incomplete combustion in auto, truck, and bus engines. More than 87 million tons pollute the air every year in the United States, and the highest concentrations are in areas of heavy traffic. Carbon monoxide replaces oxygen in the red blood cells, affects the brain and heart, and can cause fatalities.

Hydrocarbons. These gases also come from incompletely burned organic compounds in engines, as well as from industrial processes, paints, and cleaning solvents. They irritate mucous membranes and may cause lower respiratory tract problems. The biggest danger is that they often combine with other elements to form poisonous gases.

Lead. A common pollutant outdoors as well as indoors, airborne lead comes from antiknock agents in gasoline fuel and from industrial processes.

Nitrogen dioxide. Poisonous and highly reactive, this gas derives from auto and industrial emissions. It penetrates deeply into the lungs and lower respiratory tract, causing bronchial problems and, in some cases, emphysema. It also impairs immune defenses against bacterial and viral infections. Its greatest harm results from chemical interactions that produce ozone.

Ozone. This compound is a poisonous form of oxygen and the principal component of smog. It occurs as a result of chemical reactions in the atmosphere combined with sunlight—sometimes called "photochemical smog." Ozone is a powerful eye irritant, attacks the respiratory system, causing coughing and choking, impairs lung function, aggravates asthma and bronchitis, and breaks down red blood cells.

Particulates. These are actual particles of substances, such as asbestos fiber, suspended in the air. They can be microscopic or as visible as smoke and soot and are carried by the wind, primarily from industrial processes and combustion. About 7 percent of particulates come from natural, uncontrollable sources such as sand, dust, and volcanic and forest fire ash. They are largely respiratory irritants but may pass through lung membranes to other organs. Some particles contain lead, which increases their toxicity.

Sulfur dioxide. This is another gas produced mainly from the burning of sulfur-contained fuel and from industrial processes. It aggravates lung and respiratory diseases.

Some of these outdoor air impurities can be eliminated indoors by the use of activated charcoal air filters.

Appendix K
Toxins on the Job

Actors on stage: cobalt and compounds in cosmetics and cosmetics removers; theater dusts; dyes in costume fabrics

Aircraft workers: chemicals in chlorinated solvents; hydraulic fluids; lubricating oils; nitric acid; paints, thinners, and removers; plastics and resins; welding fumes and radiation

Airplane pilots and crew: formaldehyde in upholstery fabric finishes; fuel and gasoline; oils; phenols and plastics

Anesthesiologists: ether, chloroform, halothane, nitrous oxide and other volatile gases

Auto mechanics: asbestos dust; benzene as carburetor cleaner; brake fluids; carbon monoxide; ethylene glycol; gasoline; graphite; greases; kerosene; lubricating oils; paints, removers, and thinners; plastics; soldering metals; solvents

Bakers: additives in foods; ethylene dibromide as fumigant; disinfectants; flour dusts; fluorocarbons from refrigeration; fungicides; gas from ovens

Barbers and beauticians: asbestos in some dryers; chemicals in cosmetics, depilatories, disinfectants, dyes, hair preparations, nail products, perfumes, sprays

Bookbinders: aniline and cobalt in inks and dyes; benzene and toluene in glues; lead; paper dust and fumes; shellac

Brewers: carbon dioxide and monoxide; hydrogen fluoride and sulfide; sulfur dioxide are all given off during the fermentation process

Butchers and meat workers: antibiotics; detergents; fungicides, herbicides, and pesticides from grasses and plants fed to animals; hormones used to fatten animals; vinyl fumes from meat wrapping

Carpenters and construction workers: adhesives; arsenic in paint, glass, wallpaper; asbestos fibers; bleaches; coal tars; insulation; lead; oils; plastic; solvents; stains, shellac, and varnishes; wood resins

Carpet makers, layers, sellers and cleaners: adhesives; aniline dyes; bleaches; detergents; formaldehyde; fungicide; soaps; solvents; synthetic fabrics such as polyester, polyurethane, and propylene

Ceramic workers: Natural gas, acetylene; arsenic; asbestos; beryllium; cadmium; chromium; cobalt; fluoride; lead; mercury; nickel; selenium; uranium compounds; zinc

Cosmetics makers and sellers: aluminum compounds; cobalt; dyes; hexachlorophene; parabens and titanium compounds; plus thousands of chemical ingredients

Dentists and dental technicians: adhesives; anesthetic gases and Novocain; antiseptics; antibiotics; artificial scents and flavorings; mercury and cadmium in amalgam; disinfectants; oils; plastics; platinum in alloys; radiation from X rays; zinc in cement

Dock workers: all chemicals used in making package wraps; asbestos; carbon monoxide; fumigants, fungicides, and insecticides; oils; ozone; particles in dust and smoke

Dry cleaners: amyl acetate; benzene; coal tars; carbon tetrachloride; ethylene glycol; ethylene dibromide for waterproofing; formaldehyde in spot proofing and all protective finishes; methyl ethyl ketone and other solvents; naphthalene in mothproofing; perchloroethylene

Electricians and electrical workers: aluminum and all metal compounds; asphalt in adhesives; bismuth for fuses; cadmium in solder; coal tars; fluorescent lights; graphite; lead; mercury; PCBs in insulation; plastic resins; radiation in appliances; all nuclear and radiation equipment

Farmers and agricultural workers: arsenic and other poisons in fungicides, herbicides, and pesticides; chemicals in packaging materials; coloring agents; contaminants in foods from polluted water and air; ethylene gas used to accelerate ripening; synthetic fertilizer; feeds, oils and solvents in farm equipment; vinyl chloride; wax coatings on fruits

Firefighters: potent chemicals in extinguishers; auto exhaust; gases in smoke or burning material

Fishermen: chemicals from contaminated fish handled or eaten; gases from decay and wastes dumped in water; plastics in equipment; pollutants in air; preservatives in bait; sunlight; diesel fumes

Fumigators: all chemicals in fungicides, herbicides, pesticides; benzene; chloroform; formaldehyde; formic acid; hydrogen cyanide; kerosene; lindane; malathion; propane and other propellants

Furnace workers and repairmen: asbestos; butane; carbon monoxide and dioxide; coal tars; gasoline; glass fibers; infrared radiation; kerosene; lubricating oils; sulfur dioxide; zirconium

Furniture makers, upholsterers, sellers: adhesives; amyl acetate in polish; benzene; formaldehyde; formic acid in dyes and finishes; lacquer solvents; methyl alcohol; microwave radiation in veneering; pentachlorophenol as wood preservative; polyethylene in textile finishes; propylene and other synthetics in fabrics; turpentine; waxes; stains and varnishes

Fur processors: alum; bleaches; carbon tetrachloride; dyes; formaldehyde; oils; salts; sulfuric acid

Hospital workers: anesthetics; antiseptics; patients' cosmetics and medication; detergents; disinfectants; ethyl alcohol; formaldehyde; fumigants; radiation from X rays; soaps; smoke; toluene

Ink-makers and printers: adhesives; asbestos; ammonia; arsenic; benzene; carbon black; carbon tetrachloride; cornstarch; diacetone alcohol in quick-drying inks; formaldehyde; gums; hydrogen cyanide in textile and art printing; lead in wallpaper printing; naphthalene in stencil inks; paper dust; radiation; toluene; urea-formaldehyde resins; xylene in stamp pads

Jewelers: amyl acetate as lacquer solvent; lead; mercury and compounds; nitric acid; paraffin; sulfuric acid in polishes; benzene and urea in adhesives; arsenic fumes and dust from metal impurities

Leather and tanning workers: acrylonitrile in finishing; aluminum compounds; ammonia; aniline and derivatives; coal and petroleum products; formaldehyde as tanning agent; formic acid; hydrochloric acid; hydrogen cyanide; isopropyl alcohol; nitrobenzene and xylene in dyes; oxalic acid; vinyls in plastics

Milliners: adhesives; ammonium hydroxide in rubber; aniline; benzene; formaldehyde; glycerol in finishes; methyl alcohol

Miners: antimony; asbestos; arsenic; carbon dioxide and monoxide; coal tar; chromium; formaldehyde and phenols in explosives; lead; metal and rock dusts; radiation from uranium; toluene in TNT; tolylene diisocyanate in mine tunnel coating

Paint makers and users: ammonia for water-based paints; aniline; antimony; arsenic; barium and bismuth in luminous paints; benzene; cadmium; coal tar; cobalt in pigments; fluorides; graphite; mercury; nickel; phenols; sulfuric acid; titanium; toluene as thinner; uranium in uranium paints; zinc compounds

Paper manufacturers: acrylic coating; asbestos; chlorinated diphenyls and naphthalenes for carbonless paper; chromium dyes; formaldehyde finishes; hydrogen sulfide; n-butyl acetate and paraffin for coated papers; phenols for treated paper; tin and compounds in sensitized paper

Photography lab workers: ammonia in automatic film processing; hydroxide; benzyl alcohol; ethanol; ethylene glycol in film; formaldehyde; chlorine; hydrogen peroxide and other strong chemicals in developers; ketones; phenol; quinone; sulfuric acid

Plumbers: arsine; asbestos; chloroform in pipes; graphite in pipe joints; lye and other caustics; metal dusts; polyester in plastic pipe; polyurethane as caulking agent; sulfuric acid as drain cleaner; Teflon; welding and soldering fumes

Postal workers: benzene in adhesives; dyes; dusts; glycerol in stamp pads; oils; paper finishes; toluene in inks

Textile manufacturers and sellers: antimony in fire retardants; benzene and ethylene glycol as adhesives; chlorine and hydrogen peroxide as bleaches; chromates in silk-screening; formaldehyde in nonshrink, permanent press and other finishes; hydroquinone in coatings; polyesters, polyvinyls, and all synthetic fabrics; trichloroethylene as cleaner

Veterinarians: anesthetics; antibiotics; antifungal agents; antiseptics; chloroform as insecticide; glycerol as disinfectant; hexachlorophene as germicide; lindane as parasite killer; medications and urea in feed

Appendix L
Resources

The authors do not endorse these commercial programs. It is up to each individual to investigate carefully before joining any group or starting any therapy.

Alcohol (See also Detox Centers)

Alcoholics Anonymous (AA)
P.O. Box 459
Grand Central Station
New York, NY 10163
(212) 473-6200

National Clearinghouse for Alcohol Information
Box 2345
Rockville, MD 20852
(301) 468-2600

National Council on Alcoholism
733 Third Ave.
New York, NY 10017

The Grapevine (AA newsletter)
P.O. Box 1980
Grand Central Station
New York, NY 10163
(212) 686-1100

Women for Sobriety
P.O. Box 618
Quakertown, PA 18951
(215) 536-8026

Detox Centers, Clinics and Facilities

(I–Inpatient or residential facilities; O–Outpatient or nonresidential facilities.)

Beverly Detox Center–O
314 N. Harper Ave.
Los Angeles, CA 90048
(213) 655-5928

Beverly Glen Hospital–I
10361 W. Pico Blvd.
Los Angeles, CA 90064
(213) 277-5111 or 1-800-
    262-5463 for information
    and referral

Bry-Lin Hospital; Rush
Hall–I and O
Detoxification and Entry to
  Care
1263 Delaware Ave.
Buffalo, NY 14209
(716) 886-8200

CareUnit, c/o Comprehen-
sive Care Corporation–I
660 Newport Center Drive
Newport Beach, CA 92660
For info on centers all over
  the U.S., call: (800) 422-
  4427

Central Hospital–I
26 Central St.
Somerville, MA 02143
(617) 625-8900

Cumberland Heights; Drug
and Chemical Detox Unit–I
Rt. 2, River Road
Nashville, TN 37209
(615) 352-1757

Edwin Shaw Hospital–I
and O
Alcohol and Chemical De-
pendency Detox Center
1625 Flickenger Road
Akron, OH 44312
(216) 784-1271

Erie County Medical
Center–I
Substance Intervention Unit
462 Grider St.
Buffalo, NY 14215
(716) 898-3625

Eugenia Hospital, Inc.–I
The Adapt Unit; Substance
Abuse and Detox
Thomas Road
Lafayette Hills, PA 19444
(215) 836-1380

Forest Hospital–I
Hawthorne House; Drug
and Chemical Detox Center
555 Wilson Lane
Des Plaines, IL 60016
(312) 635-4300

French Hospital S.H.A.R.E.
Unit–I and O
4131 Geary Blvd.
San Francisco, CA 94118
(415) 386-1212

Garden Sullivan Hospital of
Pacific Medical Center–I
and O
2750 Geary Blvd.
San Francisco, CA 94118
(415) 921-6171

Haight Ashbury Free Medi-
cal Clinics–O
Detoxification Project
529 Clayton
San Francisco, CA 94117
(415) 621-2014

Houston International
Hospital–I
Alcohol and Drug Detox
Unit
6441 Main St.
Houston, TX 77030
(713) 795-5921

Mercy Medical Center–I
16th Ave. at Milwaukee St.
Denver, CO 80206
Adults: (303) 393-3503
Adolescents: (303) 393-
3450

Merritt Peralta Institute–I
and O
435 Hawthorne Ave.
Oakland, CA 94609
(415) 652-7000

Rochford Clinic–O
9730 Wilshire Blvd. #215
Beverly Hills, CA 90212
(213) 859-0444

Palm Beach Institute–I
1014 N. Olive Ave.
West Palm Beach, FL 33401
(305) 833-7553

St. Helena Hospital and
Health Center–I
Deer Park, CA 94576
(800) 862-7575 or (707)
963-6204

The PBI Hospital Program–I
1210 South Old Dixie
Highway
Jupiter, FL 33458
(305) 746-6602

Drugs

American Council on
Marijuana
767 Fifth Ave.
New York, NY 10022
(212) 758-6498

Cocaine HOTLINE: 800-
COCAINE for information
on treatment and resources

Cokenders
Wilbur Hot Springs Health
Sanctuary
Williams, CA 95987
(916) 473-2306

Drug Abuse Programs of
America
P.O. Box 5487
Pasadena, TX 77505
(713) 479-8440
(Write for literature about
hospital treatment)

Narcotics Anonymous
P.O. Box 459
Grand Central Station
New York, NY 10017
(212) 420-9400

Narcotics Anonymous
World Service Ofc., Inc.
16155 Wyandotte St.
Van Nuys, CA 91406

National Clearinghouse for
Drug Abuse Information
5600 Fishers Lane
Rockville, MD 20857
(301) 443-6500
(Write for free list of 3000
drug abuse treatment
centers)

National Institute on Drug Abuse
P.O. Box 2305
Rockville, MD 20852
(301) 443-4577

Straight, Inc.
P.O. Box 40052
St. Petersburg, FL 33743
(813) 345-3932
(Youth-oriented drug
   rehabilitation)

Environmental

Citizens Against Toxic Sprays
1385 Bailey Ave.
Eugene, OR 97402

Environmental Action
1346 Connecticut Avenue, N.W.
Suite 731
Washington, DC 20036
(Political action against
   toxic substances)

Environmental Defense Fund
1525 18th St., N.W.
Washington, DC 20036
(202) 833-1484
(Researchers and activists
   against toxic materials)

Environmental Illness Law Report
Editor Earon S. Davis
P.O. Box 1739
Evanston, IL 60204–1369

Environmental Protection Agency
401 M St., S.W.
Washington, D.C. 20460
(202) 755-2700

EPA Public Inquiry Center:
(202) 755-0707
EPA switchboard: (202) 382-2090

Friends of the Earth
1045 Sansome St.
San Francisco, CA 94111
(415) 433-7373

Human Ecology Action League (HEAL)
Attn. Earon S. Davis
P.O. Box 1369
Evanston, IL 60204–1369
(312) 864-0995

National Clean Air Coalition
620 C St., S.E.
Washington, DC 20003
(202) 543-0305

National Institute of Environmental Health Sciences
P.O. Box 12233
Research Triangle Park, NC 27709
(919) 541-3345
(Conducts research on environmental hazards)

National Institute for Occupational Safety and Health
Parklawn Bldg.
5600 Fishers Lane
Rockville, MD 20857
(301) 443-2140

National Solid Waste Management Association
1120 Connecticut Ave., N.W.
Washington, DC 20036
(202) 659-4613
(Supplies information on waste disposal)

Northwest Coalition for Alternatives to Pesticides
Box 375
Eugene, OR 97440

Occupational Safety and Health Administration
200 Constitution Ave., N.W.
Washington, DC 20210
(202) 523-9362
(Deals with controlling toxins in workplace)

Sierra Club
530 Bush St.
San Francisco, CA 94108
(415) 981-8634

U.S. Geological Survey
Water Resources Division
National Water Data Exchange
Reston, VA 22092
(703) 860-7444 or (703) 860-6031
(Supplies information on groundwater)

General

Center for Science in the Public Interest
1755 S Street, N.W.
Washington, DC 20009
(202) 332-9110
(Activists against toxic chemicals in food)

Clearinghouse for Occupational Safety and Health Information
Robert A. Taft Laboratories
4676 Columbia Parkway
Cincinnati, OH 45226

Consumer Product Safety Commission HOTLINE:
(800) 638-8326

Consumer Information Center
Department L
Pueblo, CO 81009
(Write for consumer information catalog)

Federation of Homemakers
Box 5571
Arlington, VA 22205

Food Safety and Inspection
Service
Meat & Poultry Hotline
U.S. Department of
    Agriculture
Washington, DC 20250

National Center for Toxicological Research
General Services, Food &
    Drug Administration De-
    partment of Health & Hu-
    man Services
Jefferson, AR 72079
(501) 541-4344

National Institute of Mental
Health
U.S. Department of Health
    and Human Services
5600 Fishers Lane
Rockville, MD 20857
(301) 245-6296
(Write for names of groups
    interested in your
    problem)

National Self-Help
Clearinghouse
33 W. 42nd St.
Rm. 1206-A
New York, NY 10036
(Write for information on
    forming your own group)

Nutrition for Optimal
Health Association
Box 380
Winnetka, IL 60093

Self-Help Center
1600 Dodge Ave.
Suite S-122
Evanston, IL 60201
(Write for information on
    forming your own group
    p. 245A.)

Medical

National Cancer Institute
9000 Rockville Pike
Bethesda, MD 20014
(301) 496-5615
(Conducts cancer research)

Pills Anonymous
Box 459
Grand Central Station
New York, NY 10017
(212) 874-0700

Society for Clinical Ecology
Judy Howard, Administra-
    tive Assistant
P.O. Box 16106
Denver, CO 80216
(303) 622-9755
(Write for name of clinical
    ecologists in your area)

U.S. Pharmacopeial
Convention
Drug Information Division
12601 Twinbrook Parkway
Rockville, MD 20852

Smoking

Action on Smoking and
Health
2000 H St., N.W.
Washington, DC 20006
(202) 659-4310
(Political action for rights of
nonsmokers)

American Cancer Society
(check telephone listings for
local chapter)
777 Third Ave.
New York, NY 10017
(212) 371-2900

American Heart Association
(check telephone listings for
local chapter)
7320 Greenville Ave.
Dallas, TX 75231
(214) 750-5300

American Lung Association
(check telephone listings for
local chapter)
1740 Broadway
New York, NY 10019
(212) 245-8000

Clearing the Air
Public Health Service
Rockville, MD 20857
(Write for free booklet or
call 1-800-4-CANCER)

Five-Day Plan to Stop
Smoking
Seventh-Day Adventist
Church
Narcotics Education
Division
6840 Eastern Ave., N.W.
Washington, DC 20012
(202) 723-0800

Mayo Clinic
200 First St., S.W.
Rochester, MN 55909
(507) 282-2511

Office on Smoking and
Health
158 Park Building
5600 Fishers Lane
Rockville, MD 20857
(301) 443-5287

Schick Laboratories
1901 Avenue of the Stars,
Suite 1530
Los Angeles, CA 90067

SmokEnders
3708 Mt. Diablo Blvd.
Lafayette, CA 94549
(800) 227-2334

Recommended Reading

American Society of Hospital Pharmacists. *Consumer Drug Digest*. New York: Facts on File, 1982.

Bargmann, Eve; Wolfe, Sidney; and Levin, Joan. Public Citizen Health Research Group. *Stopping Valium*. New York: Warner, 1982.

Baron, Jason D. *Kids & Drugs*. New York: Putnam, 1984.

Bell, Iris R. *Clinical Ecology*. Bolinas, Calif.: Common Knowledge Press, 1982.

Benowicz, Robert J. *Non-Prescription Drugs and Their Side Effects*. New York: Berkley, 1982.

Boyle, Robert H. *Acid Rain*. New York: Nick Lyons Books, 1983.

Brecher, Edward M., and the Editors of Consumer Reports. *Licit & Illicit Drugs*. Boston: Little, Brown, 1972.

Brown, Edmund Jr.; Lytle, Alice A.; and Spohn, Richard B. *Clean Your Room!* California: Department of Consumer Affairs, 1982.

Calabrese, Edward J., and Dorsey, Michael W. *Healthy Living in an Unhealthy World*. New York: Simon and Schuster, 1984.

Casarett, Louis J., and Doull, John. *Toxicology*. New York: Macmillan, 1980.

Crook, William G. *The Yeast Connection*. Jackson, Tenn: Professional Books, 1983.

Cummings, Stephen, and Ullman, Dana. *Everybody's Guide to Homeopathic Medicines*. Los Angeles: Tarcher, 1984.

Dadd, Debra Lynn. *Nontoxic & Natural: How to Avoid Dangerous Products and Make Healthy Ones*. Los Angeles: Tarcher, 1984.

Dickey, Lawrence D. *Clinical Ecology.* Springfield, Ill.: Charles C. Thomas, 1976.

Dreisbach, Robert H. *Handbook of Poisoning.* California: Lange, 1983.

Dufty, William. *Sugar Blues.* New York: Warner, 1976.

Editors of Consumer Report Books. *The Medicine Show.* New York: Consumers Union, 1980.

Editors of Consumer Report Books. *The Product Safety Book.* New York: Consumer Federation of America, Dutton, 1983.

Edmonds, Alan; Soyka, Fred. *The Ion Effect.* New York: Dutton, 1977.

Epstein, Samuel; Brown, Lester O.; and Pope, Carl. *Hazardous Waste in America.* San Francisco: Sierra Club Books, 1982.

Geisinger, David L. *Kicking It.* New York: Grove, 1978.

Gelb, Harold. *Killing Pain Without Prescription.* New York: Barnes & Noble, 1982.

Giannini, A. James; Slaby, Andrew E.; and Giannini, Matthew C. *Handbook of Overdose and Detoxification Emergencies.* New York: Medical Examination Publishing Co., 1982.

Glabman, Sheldon, and Freese, Arthur S. *Your Kidneys, Their Care and Their Cure.* New York: Dutton, 1976.

Gross, Leonard. *How Much Is Too Much?* New York: Random House, 1983.

Hill, Amelia Nathan. *Against the Unsuspected Enemy.* W. Sussex: New Horizon, 1980.

Hunter, Beatrice Trum. *Additives Book.* New Canaan, Conn.: Keats, 1980.

Hunter, Beatrice Trum. *Great Nutrition Robbery.* New York: Scribner's, 1978.

Hunter, Beatrice Trum. *How Safe Is Food in Your Kitchen?* New York: Scribner's, 1981.

Hunter, Beatrice Trum. *The Sugar Trap and How to Avoid It.* Boston: Houghton Mifflin, 1982.

Hunter, Beatrice Trum. *The Mirage of Safety: Food Additives and Federal Policy.* Brattleboro, Vermont. Stephen Greene, 1982.

Kissin, Benjamin; Lowinson, Joyce; and Millman, Robert. *Recent Developments in Chemotherapy of Narcotic Addiction.* New York: Academy of Sciences, 1978.

Levin, Alan S., and Zellerbach, Merla. *The Type 1/Type 2 Allergy Relief Program.* Los Angeles: Tarcher, 1983.

Livingston-Wheeler, Virginia, and Addeo, Edmond G. *The Conquest of Cancer.* New York: Watts, 1984.

Lipske, Michael. *Chemical Additives in Booze.* Washington, D.C.: Center for Science in the Public Interest, 1982.

Mackarness, Richard. *Chemical Victims.* London: Pan Books, 1980.

Makower, Joel. *Office Hazards.* Washington, D.C.: Tilden, 1981.

Mandell, Marshall; Scanlon, Lynne Waller. *5-Day Allergy Relief System.* New York: Pocket Books, 1979.

Mindell, Earl. *Earl Mindell's Pill Bible.* New York: Bantam, 1984.

O'Banion, Dan. *An Ecological and Nutritional Approach to Behavioral Medicine.* Springfield, Ill.: Charles C. Thomas, 1981.

O'Banion, Dan. *The Ecological and Nutritional Treatment of Health Disorders.* Springfield, Ill.: Charles C. Thomas, 1981.

Ott, John N. *Health and Light.* New York: Simon and Schuster, 1973.

Panos, Maesimund B., and Heimlich, Jane. *Homeopathic Medicine at Home.* Los Angeles: Tarcher, 1980.

Pfeiffer, Guy, and Nickel, Casimir. *The Household Environment and Chronic Illness.* Springfield, Ill.: Charles C. Thomas, 1980.

Randolph, Theron G., and Moss, Ralph W. *An Alternative Approach to Allergies.* New York: Lippincott & Crowell, 1980.

Randolph, Theron G. *Human Ecology and Susceptibility to the Chemical Environment.* Springfield, Ill.: Charles C. Thomas, 1962.

Rapp, Doris J. *Allergies and the Hyperactive Child.* New York: Simon and Schuster, 1979.

Rohe, Fred. *The Complete Book of Natural Foods.* Colorado: Shambhala, 1983.

Roppere, Vickey. *The Allergy Problem.* Miami: Meded Publisher Inc., 1981.

Rosenblatt, Seymour, and Dodson, Reynolds. *Beyond Valium.* New York: Putnam, 1981.

Ryan, Regina Sara, and Travis, John W. *Wellness Workbook.* Berkeley, Calif.: Ten Speed Press, 1981.

Samuels, Mike, and Bennett, Hal Zina. *Well Body, Well Earth.* San Francisco: Sierra Club Books, 1983.

Schauss, Alexander. *Diet, Crime and Delinquency.* Berkeley, Calif.: Parker House, 1980.

Shakman, Robert A. *Poison-Proof Your Body.* New York: Van Nostrand Reinhold, 1982.

Small, Barbara, and Small, Bruce. *Sunny Hill.* Ontario: Small & Associates, 1980.

Small, Bruce. *The Susceptibility Report.* Longueuil P.Q., Canada: Deco Plan Inc., 1982.

Stellman, Jeanne M., and Daum, Susan M. *Work Is Dangerous to Your Health.* New York: Vintage Books, 1973.

Vander, Arthur J. *Nutrition, Stress and Toxic Chemicals.* Ann Arbor: University of Michigan Press, 1981.

Vogler, Roger E., and Bartz, Wayne R. *The Better Way to Drink.* New York: Simon and Schuster, 1982.

Weiner, Michael A. *Getting Off Cocaine.* New York: Avon, 1984.

Weiner, Michael A., and Goss, Kathleen. *Nutrition Against Aging.* New York: Bantam, 1983.

Wetherall, Charles F. *Kicking the Coffee Habit.* Minnesota: Wetherall, 1981.

Zamm, Alfred V., and Gannon, Robert. *Why Your House May Endanger Your Health.* New York: Simon and Schuster, 1982.

Health-Oriented Cookbooks and Books with Special Diets

Golos, Natalie; Golbitz, Frances Golos; and Leighton, Frances Spatz. *Coping With Your Allergies.* New York: Simon and Schuster, 1979.

Hewitt, Jean. *The New York Times New Natural Foods Cookbook.* New York: Avon, 1983.

Katzen, Mollie. *Moosewood Cookbook.* Berkeley, Calif.: Ten Speed Press, 1977.

Kushi, Michio, and Jack, Alex. *The Cancer Prevention Diet.* New York: St. Martin's Press, 1983.

Ogle, Jane. *The Stop Smoking Diet.* New York: Evans, 1981.

Pritikin, Nathan, and McGrady, Patrick M. Jr. *The Pritikin Program For Diet and Exercise.* New York: Bantam, 1983.

Robertson, Laurel, and Flinders, Carol, and Godfrey, Bronwen. *Laurel's Kitchen.* New York: Bantam, 1982.

Shattuck, Ruth R. *The Allergy Cookbook.* New York: Plume, 1984.

Smith, Lendon. *Feed Yourself Right.* New York: McGraw-Hill, 1983.

Index

Premenstrual syndrome, 132
Preservatives, 42, 45
Proteins, 65–66
Psychological tests, 83

Quality of life, 156

Radiation, 40, 176. *See also*
    X rays
Randolph, Theron, 120, 130
Rashes, 16–17, 36
Recipes, Detox Diet
    Green Sauce, 73
    Parsley-Chicken Soup, 73
    Tomato-Beef Broth, 72
    Waldorf Salad, 73
Recreational drugs, 52–55. *See
    also* Cocaine; Marijuana
    alternatives, 95–99
    in pregnancy, 177
Relaxation, 76–78
Religious cult dogma, 3
Repace, James L., 51
Respirators, 61–62
Respiratory system, 16, 23,
    24–25, 76, 108
Rheumatoid arthritis, 8
Roe, Bob, 123
Rosenblatt, Seymour, 151
Rotation diet, 68, 131
Runner's high, 76

Saccharin, 45
Salt, 44–45, 67, 70, 77
Sassafras, 42, 97
Scents. *See* Odors
Schnare, David W., 136
Schweiker, Richard, 156
Sebaceous glands, 23, 30
Selenium, 14, 65

Self-purification, 3–4
Senna, 97
Sex drive, 13
Skin, 15, 23, 30, 108. *See also*
    Contact reactions;
    Contactants
Sleep, 12, 151
Smell, sense of, 14
Smith, David E., 126, 128
Smith, Lendon, 133, 134
Smith, Michael O., 114
Smoking. *See* Nicotine
Snuff, 98–99
Sodas, 45
Street, J. C., 122
Stress, and hormones, 76
Sugar
    addiction, 102, 131–133
    detoxification program,
        133–134
    sources, 41, 43
    toxic effects, 43–44, 67
    withdrawal symptoms,
        106–110
Sugar Blues (Dufty), 131, 134
Sugar Detox Procedure, 133–134
Support groups, 77–78, 128
Susceptibility to toxic re-
    sponse, 8–10
Sweating, 18, 23, 30, 75–76,
    83
Swimming, 75
Symptoms. *See* Toxicity symp-
    toms; Withdrawal
    symptoms
Synthetics, 58–59

Tea. *See* Herbal teas
Technological advances,
    147–157

Phyllis Saifer, M.D., M.P.H., is in private medical practice in Berkeley, California and has organized the San Francisco Bay Area chapter of the Environmental Illness Association. She is a frequent lecturer on environmental aspects of health and disease, and currently serves as national secretary and as editor of "SCE Scene," the newsletter of the Society for Clinical Ecology.

Merla Zellerbach is a twenty-two year columnist for the San Francisco Chronicle, a frequent contributor to national magazines, the author of *Love in a Dark House*, which deals with society's treatment of the mentally ill, and co-author with Dr. Alan Scott Levin of *The Type 1/Type 2 Allergy Relief Program*.